Structural
Analysis

Structural
Analysis

ALEXANDER CHAJES

Department of Civil Engineering
University of Massachusetts

PRENTICE-HALL, INC., ENGLEWOOD CLIFFS, N. J. 07632

Library of Congress Cataloging in Publication Data

CHAJES, ALEXANDER.   (date)
   Structural analysis.

   (Prentice-Hall civil engineering and engineer-
ing mechanics series)
   Includes index.
   1. Structures, Theory of. I. Title. II. Series.
TA645.C37 1982   624.1′71   82–12254
ISBN 0–13–853408–X

PRENTICE-HALL CIVIL ENGINEERING AND ENGINEERING MECHANICS SERIES
W. J. Hall, Editor

Editorial production/supervision and interior design:
Gretchen K. Chenenko
Cover design: *Karolina Harris*
Cover photo: Citicorp Center, N.Y.C., N.Y.
Courtesy of Hugh Stubbins and Associates, Inc.
Architects. Photograph: Edward Jacoby/APG.)
Art production: *Marie Dobish*
Manufacturing buyer: *Anthony Caruso*

Printed in the United States of America

10 9 8 7 6 5

ISBN 0-13-853408-X

Prentice-Hall International, Inc., *London*
Prentice-Hall of Australia Pty. Limited, *Sydney*
Editora Prentice-Hall do Brasil, Ltda., *Rio de Janeiro*
Prentice-Hall Canada Inc., *Toronto*
Prentice-Hall of India Private Limited, *New Delhi*
Prentice-Hall of Japan, Inc., *Tokyo*
Prentice-Hall of Southeast Asia Pte. Ltd., *Singapore*
Whitehall Books Limited, *Wellington, New Zealand*

*To Diane, Susan,
and Michael*

Contents

Preface   *xiii*

1   Introduction   *1*

1.1   Structural Engineering   *2*
1.2   Structural Form   *2*
1.3   Materials   *9*
1.4   Loads   *12*
1.5   History of Structural Engineering   *15*

2   Calculation of Reactions   *19*

2.1   Introduction   *20*
2.2   Equations of Equilibrium   *20*
2.3   Types of Supports and Restraints   *21*
2.4   Stability and Determinacy   *22*
2.5   Calculation of Reactions—Simple Structures   *25*
2.6   Calculation of Reactions—Compound Structures   *27*
        Problems   *31*

3 Plane Trusses

35

3.1 Introduction 36
3.2 Simplifying Assumptions 36
3.3 Basic Concepts 38
3.4 Method of Joints 39
3.5 Method of Sections 45
3.6 Methods of Joints and Sections Combined 48
3.7 Stability and Determinacy of Trusses 51
3.8 Computer Solution 53
    Problems 59

4 Space Trusses

65

4.1 Introduction 66
4.2 Calculation of Reactions 66
4.3 Calculation of Member Forces 67
    Problems 74

5 Shear and Moment Diagrams for Beams and Frames

77

5.1 Introduction 78
5.2 Internal Forces 79
5.3 Sign Convention 80
5.4 Shear and Bending-Moment Diagrams by the Method of Sections 80
5.5 Relationships between Load, Shear, and Bending Moment 83
5.6 Construction of Shear and Moment Diagrams 85
5.7 Shear and Moment Diagrams of Determinate, Rigid Frames 91
    Problems 98

6 Deflections: Differential Equation

103

6.1 Introduction 104
6.2 Elastic Force-Deformation Relationships 104
6.3 Direct Integration 106
6.4 Moment-Area Theorems 110
6.5 Application of the Moment-Area Method 112
6.6 Conjugate-Beam Method 120
6.7 Application of the Conjugate-Beam Method 124
    Problems 128

7 Deflections: Energy Methods 135

7.1 Introduction *136*
7.2 Principle of Conservation of Energy *136*
7.3 Method of Real Work *138*
7.4 Virtual Work *139*
7.5 Application of Virtual Work to Beams and Frames *142*
7.6 Deflection of Trusses Using Virtual Work *149*
  Problems *152*

8 Influence Lines 159

8.1 Introduction *160*
8.2 Influence Lines Defined *160*
8.3 Influence Lines for Beams *161*
8.4 Influence Lines for Trusses *167*
8.5 Uses of Influence Lines *172*
  Problems *175*

9 Arches and Cables 179

9.1 Arches *180*
9.2 Cables *183*
  Problems *185*

10 Indeterminate Structures: Introduction 187

11 Method of Consistent Deformations 191

11.1 Introduction *192*
11.2 Basic Principles *193*
11.3 Application of Consistent Deformations to Structures with One Redundant *195*
11.4 Support Settlement *204*
11.5 Structures with Several Redundants *206*
11.6 Structures with Internal Redundants *212*
  Problems *214*

12 Method of Least Work 219

12.1 Introduction *220*
12.2 Derivation of Castigliano's Theorems *220*
12.3 Application of the Method of Least Work *223*
  Problems *232*

13   **Slope-Deflection Method**   *237*

   13.1   Introduction   *238*
   13.2   Derivation of the Slope-Deflection Equation   *238*
   13.3   Alternate Derivation of Slope-Deflection Equation   *242*
   13.4   Application of the Slope-Deflection Method   *244*
          Problems   *254*

14   **Moment-Distribution Method**   *257*

   14.1   Introduction   *258*
   14.2   Basic Concepts   *260*
   14.3   Application of Moment Distribution to Structures
          Without Joint Translations   *263*
   14.4   Structures with Joint Translations   *271*
          Problems   *275*

15   **Matrix Flexibility Method**   *277*

   15.1   Introduction   *278*
   15.2   Flexibility Matrix   *278*
   15.3   Formation of the Structure-Flexibility Matrix
          from Element-Flexibility Matrices   *281*
   15.4   Analysis of Indeterminate Structures   *289*
          Problems   *297*

16   **Matrix Stiffness Method**   *301*

   16.1   Introduction   *302*
   16.2   Stiffness Matrix   *302*
   16.3   Element Stiffness Matrix   *303*
   16.4   Formation of the Structure-Stiffness Matrix
          from Element-Stiffness Matrices   *307*
   16.5   The Direct Stiffness Method of Forming
          the Structure-Stiffness Matrix   *310*
   16.6   Analysis of Trusses by the Direct Stiffness Method   *314*
   16.7   Analysis of Flexural Structures by Direct Stiffness Method   *322*
          Problems   *328*

Appendix: Matrix Algebra   *333*

A.1   Definitions   *333*
A.2   Types of Matrices   *334*
A.3   Equality, Addition, and Subtraction   *335*
A.4   Multiplication   *335*
A.5   Transpose of a Matrix   *337*
A.6   Determinants   *337*
A.7   Inverse of a Matrix   *339*
A.8   Partitioning of Matrices   *342*

Selected Answers to Even-Numbered Problems   *345*

Index   *349*

Preface

The aim of this book is to present the fundamentals of structural analysis and to serve as a textbook for one or more courses in the subject. The material covered by the book can be subdivided into three parts. The first part, which includes Chapters 1 through 9, deals with the analysis of simple, determinate structures. This section covers the analysis of trusses, beams, frames, arches, and cables as well as methods for calculating deflections and the use of influence lines for moving loads. The second part, which includes Chapters 10 through 14, presents the classical methods of analyzing indeterminate structures. Two force methods, the method of consistent deformation and the method of least work, and two deformation methods, the slope-deflection method and moment distribution, are included. The last part of the book, Chapters 15 and 16, deals with matrix analysis of structures. Both the flexibility method and the stiffness method are presented.

The first two sections of the book, dealing with the analysis of determinate and indeterminate structures by classical methods, are presented without recourse to matrix algebra. It is felt that the inclusion of matrix algebra in this part of the book would only detract from the principal objective—namely, to introduce the student to the fundamentals of structural analysis. Once the student has mastered the basic principles presented in the first part of the book, he or she should have no difficulty with the concepts of matrix analysis covered in the latter part of the book. Matrix algebra is a relatively simple and straightforward subject, which in the author's opinion does not warrant an extended period of familiarization. However, the Appendix includes a short, concentrated review of the material which may be presented prior to introducing matrix analysis.

Although no matrix algebra is employed in the first two parts of the book, a section is included at the end of Chapter 3 that allows the student to make use of electronic computers if he or she so desires.

Depending on the amount of time available to the instructor, either one or both of the two chapters on matrix analysis can be covered. The flexibility method, presented in Chapter 15, has more in common with the classical methods of analysis than does the stiffness method. As a consequence, the concepts involved in the flexibility method are relatively easy to grasp, and the method provides a smooth transition from the classical methods to matrix analysis. However, the stiffness method, because it can be readily automated, is the matrix method most commonly used in actual practice. Unfortunately, the same mathematical elegance that leads to the advantages of the stiffness method also tends to obscure somewhat the basic principles of structural analysis that are the basis of the method.

If sufficient time is available, it is desirable to cover both the flexibility method and the stiffness method in the order presented in the book. However, if this is not possible, the book is written so that either of the two methods can be studied without the other.

The author wishes to express his gratitude to Anne Storozuk and Michaline Ilnicky, who typed the manuscript, and to David Brockelbank, Eric Chen, and Jim Churchill, who checked much of the numerical work. The author is also grateful to his family for their patience and understanding.

ALEXANDER CHAJES
Amherst, Massachusetts

1

Introduction

Mackinac Straits Bridge,
St. Ignace, Mich.
*(Courtesy of American
Bridge Division, U.S. Steel
Corporation.)*

1.1 STRUCTURAL ENGINEERING

The purpose of this book is to introduce the student to the principles of structural analysis. However, structural analysis is part of the larger subject of structural engineering, and it is with the latter that we shall begin.

Structural engineers may find themselves involved in the design of a large variety of projects including buildings, bridges, dams, ships, airplanes, sports stadiums, and nuclear power plants. Although each of these projects has a different purpose, they all have some common characteristics. To begin with, they are all built to satisfy human needs. Buildings are erected to provide shelter, bridges to facilitate the movement of people across rivers, dams to contain large quantities of water, and airplanes to provide rapid transportation. Furthermore, in each of these examples, the structure while carrying out its intended function is subjected to various loads. The structure must always support the load due to its own weight. In addition, it must resist loads such as the weight of snow on the roof of a building or the dome of a sports stadium, the weight of people and vehicles on a bridge, the pressure of water on the sides of a dam, and the force of the wind on the wings of an airplane. It is the responsibility of the structural engineer to see to it that these loads can be resisted by the structure both safely and economically.

The structural design of a project can usually be broken down into the following four steps:

1. Selection of the type of structural form to be used and the material out of which the structure is to be made.
2. Determination of the external loads that can be expected to act on the structure.
3. Calculation of the stresses and deformations that are produced in the individual members of the structure by the external loads.
4. Determination of the size of individual members so that existing stresses and deformations do not exceed allowable values for the given material.

In this book we are primarily concerned with step 3, the analysis of structural behavior. However, all four steps in the design process are interrelated, and we will therefore devote the remainder of this chapter to aspects of the design process other than analysis.

1.2 STRUCTURAL FORM

It has been pointed out that a structure, while carrying out its intended function, is subjected to various loads that it must resist without either collapsing or deforming excessively. To put it very simply, resisting a load means transferring the load from the point at which it is applied to some type of rigid foundation. For example, the frame of a building transmits the loads due to wind and snow, which are applied to the sides and the roof of the building, to the foundation in the ground. Similarly, a

2

bridge transmits the weights of vehicles applied along its span to the abutments at the ends of the span. The manner in which a structure carries out this transmission of loads depends largely on its geometrical shape or form. Although there exist an endless number of different shapes and forms that a given structure can assume, it can be shown that these are to a great extent simply variations of a handful of basic forms.

The most important characteristic of a structural form is the type of stress it develops in resisting applied loads, i.e., uniform axial stress, bending stresses, or a combination of the two. Accordingly, structural forms can be divided into the following categories.

A. Cables.   Probably the simplest way of transmitting a load P laterally, to two supports A and B, is by means of a cable as shown in Fig. 1.1a. In this system the load is maintained in equilibrium solely by means of uniform tension stresses in the cable. Since the applied load is balanced by the vertical components of the cable forces, the cable must have a sag in order to resist the load. The larger the sag, the smaller will be the tension that the load induces in the cable. For a cable system to function properly, the supports must be able to resist the pull of the cables.

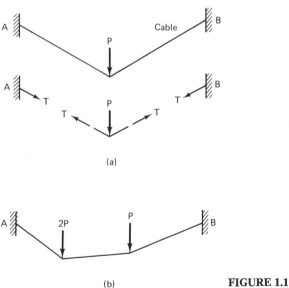

(a)

(b)   **FIGURE 1.1**

If the position of the load changes, or if several loads act simultaneously, the geometry of the cable changes accordingly, as shown in Fig. 1.1b. This ability of the cable to adjust its shape to different loading conditions is an extremely important characteristic of the cable. It allows the cable always to assume that shape for which only tension stresses are needed to resist the applied load. The shape a cable assumes when subjected to a given loading is called the *funicular curve* for that loading. For example, the funicular curve for a load that is uniformly distributed in the horizontal direction is a parabola. The load that is applied to the cables of a suspension bridge is very nearly uniform in the horizontal direction, and the cables accordingly assume shapes that are approximately parabolic. If the load is uniformly distributed along the length of the cable, the cable will take on the form of a catenary. A cable of uniform cross section subjected only to its own weight will assume the form of a catenary.

Probably the most familiar example of a cable structure is the suspension bridge. As shown in Fig. 1.2, the basic structural system consists of a horizontal deck or roadway, two vertical towers, and a pair of cables draped over the towers. The roadway is suspended from the cables by a series of vertical rods called *hangers*. The cables in turn are held up by the towers and anchored at their extremities by large concrete foundations. The entire weight of the deck, plus that of any vehicles thereon, is thus transferred to the towers and the anchorages at the ends of the cables by tension stresses in the cables.

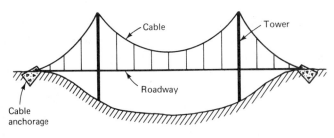

FIGURE 1.2

B. Arches. If the cable in Fig. 1.1a is inverted and replaced by two rigid bars, we have the simple arch shown in Fig. 1.3. In this system the load P is transmitted to the supports at A and C by means of uniform compression stresses in each of the struts. As was the case with a cable, it is necessary in the design of an arch to ensure that the foundations are strong enough to support the system adequately. In contrast to the cable, which pulls on the supports, the arch exerts outward thrusts on its supports.

For any single loading condition it is possible to design an arch that can resist that loading by developing only axial compression. However, most arches are sub-

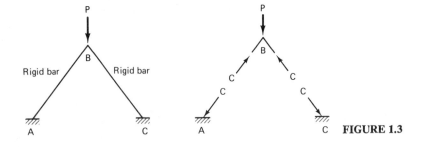

FIGURE 1.3

jected to several different loading conditions during their life, and since an arch, unlike a cable, is a rigid structure that cannot change its shape when subjected to different loadings, it will develop bending stresses in addition to axial compression. In designing an arch it is customary to choose its shape so that the predominant loading will produce a uniform compression stress. The bending that then results from other loads will be minimal. For example, the main loading of a bridge is its own weight, and since this load roughly corresponds to a uniformly distributed load, most arch bridges are approximately parabolic in shape.

C. Trusses.   The lower ends of the two slanting members of the arch in Fig. 1.3 are prevented from moving outward by the supports at *A* and *C*. Alternatively, the lower ends of the structure can be kept from spreading apart by connecting them with a third bar as indicated in Fig. 1.4. If this is done, we have, instead of an arch, a *truss*. The slanting members are still in compression, but there is now an additional member, the horizontal bar, that is in tension.

In general, a truss is a structure composed of axially stressed bars, some of which are in tension and some in compression, and in which the bars are arranged to form one or more triangles.

If the three-bar system we are considering is loaded at point *D* along the horizontal bar instead of at point *B* as before, we no longer have a truss. The system may still look like a truss, but it can no longer behave like one. The load is now transferred laterally to the supports not by axial stresses in the bars, which is the way a truss behaves, but by bending of member *AC*. To cause the system, loaded at *D*, to behave as a truss, it is necessary to add a fourth bar, connecting points *D* and *B*, as shown in Fig. 1.5. The applied load can now be transferred from *D* to *B* by tension in member

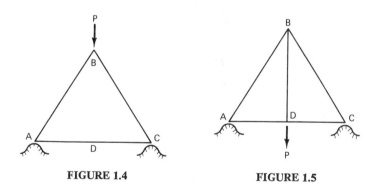

FIGURE 1.4   **FIGURE 1.5**

DB and from *B* down to the supports at *A* and *C* by compression in bars *AB* and *BC*. As before, member *ADC* prevents the lower ends of bars *AB* and *BC* from spreading and is stressed in tension.

By combining several simple triangular trusses, as shown in Fig. 1.6, it is possible to obtain structures capable of spanning relatively large distances.

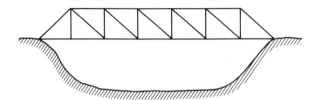

FIGURE 1.6

D. Beams. The beam shown in Fig. 1.7 transfers the load P laterally to the supports at A and B by bending. A bent member takes on a curved shape, causing the fibers along the concave edge to be shortened and in compression, and the fibers along the convex edge to be lengthened and in tension. An advantage of the beam over the arch and the cable is that it requires only vertical support at its ends. Unlike the arch and the cable, it does not require any horizontal restraint. The abutments on which the ends of a beam rest are therefore easier to design than those required by the arch and the cable. In this respect the truss is similar to the beam. It also does not require horizontal restraints at its ends.

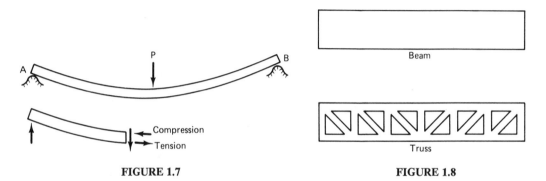

FIGURE 1.7   **FIGURE 1.8**

Another important advantage of the beam is its compactness. A beam usually occupies less vertical space than any of the structural forms considered so far. This attribute of the beam makes it especially attractive for supporting floors and ceilings in multistory buildings where vertical space is limited.

A disadvantage of the beam is that it sometimes uses material less economically than other structural systems. In an axially loaded member such as a cable or the bar of a truss, the entire cross-sectional area of the member is subjected to a uniform stress. Thus every element of the section can be stressed up to the allowable strength of the material. By comparison the magnitude of the axial stress varies from a maximum at the upper and lower edges of a beam to zero at the center of the cross section. As a consequence, fibers near the center of the beam are stressed relatively slightly when the outer fibers are working near their maximum capacity. The fact that a truss could be obtained by removing much of the material of a beam, as shown in Fig. 1.8, demonstrates that a beam uses material less efficiently than a truss.

E. Surfaces—membranes, plates, and shells. Each of the structural elements considered up to now transfers loads in one direction only. By comparison,

6

FIGURE 1.9

surfaces such as membranes, plates, and shells are two-dimensional structures, which transfer loads in two directions.

Membranes are thin sheets of material that resist applied loads by developing tension stresses in much the same manner that cables do. The tent shown in Fig. 1.9 is an example of a membrane structure. Sails, balloons, and parachutes are other examples of membrane structures. Ordinarily, membranes can only resist loads that produce tension stresses in the surface. However, it is possible by prestressing a membrane to give it the capacity to resist loads that produce compression as well. For example, the walls of an automobile tire are able to support the weight of the automobile because of the internal pressure of the tire. Similarly, domes made of relatively thin material can support their own weight if they are prestressed by internal pressure. In both these cases loads that would ordinarily induce compression in the surface only reduce the tension that has been produced by prestressing the surface.

Unlike membranes, plates and shells are rigid surfaces that can develop bending stresses as well as axial tension and compression stresses. *Plates* are flat surfaces that transfer loads by bending in a manner similar to beams. In fact, a plate can be thought of as a grid made up of beams spanning in two perpendicular directions as shown in Fig. 1.10. Assuming that the plate is supported along its four edges, a load P can be transferred to these supports by bending in two perpendicular directions.

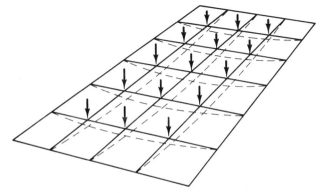

FIGURE 1.10

Like a plate, a *shell* is a rigid surface that transfers loads in two directions. The primary difference between a plate and a shell is that the shell has curvature whereas the plate does not. As a consequence, a shell is able to resist applied loads by developing axial stresses, something a flat plate is not able to do. In the curved surface shown in Fig. 1.11 the applied load P is resisted by the vertical components of the compression forces developed in the shell. By comparison, shear stresses, which produce bending, resist the applied load in a flat plate.

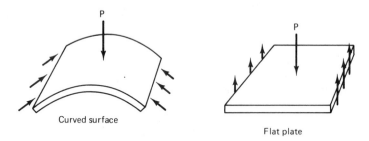

Curved surface

Flat plate

FIGURE 1.11

If a plate can be said to be a two-dimensional extension of a beam, then a shell can be thought of as a two-dimensional extension of an arch. Thus the cylindrical barrel shell and the dome in Fig. 1.12 can be formed, respectively, by placing several arches side by side and by rotating an arch about a vertical axis through its apex.

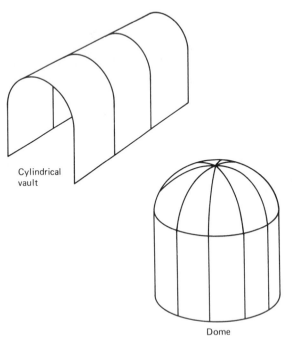

Cylindrical vault

Dome   **FIGURE 1.12**

1.3 MATERIALS

To a very large degree, the capacity of a structure to resist external loads without collapsing or deforming excessively depends on the material out of which it is made. The properties of a material that most interest the structural engineer are its strength and its deformation characteristics.

The *strength* of a material is a measure of its ability to resist stress without fracturing. Most important is the capacity of the material to resist tension and compression stress. Steel is probably the best generally available structural material because of its ability to resist equally well high tension and compression stresses without fracturing. By comparison, stone and concrete are fairly strong in compression but relatively weak in tension. Whereas structural steel can safely withstand tension and compression stresses of approximately 30 ksi (207 MN/m²), concrete can resist 3 to 7 ksi (21 to 48 MN/m²) in compression but only about 0.5 ksi (3.5 MN/m²) in tension.

Since a structure must always support its own weight, the strength-to-weight ratio of a material is of considerable importance. Wood, for example, has a strength along the grain only one tenth that of structural steel. However, the unit weight of wood is also only ten percent of the unit weight of steel. Per pound of material, wood is thus able to support as much load as steel. Because a high strength-to-weight ratio is of paramount importance in airplanes, aluminum alloys are used in their construction instead of steel. These alloys, although they are more expensive than steel, have about the same strength as structural steel but weigh only one third as much.

For a material to be suitable as a structural component, it must in addition to possessing sufficient strength have acceptable *deformation characteristics*. When subjected to stresses whose magnitudes are comparable to those expected to occur during the everyday life of the structure, the material should not deform excessively. In other words, the material must be relatively stiff. For this reason rubber, which is very flexible, is not a satisfactory structural material. Excessive deformations are objectionable because they tend to impair the proper functioning of the structure. For example, large deformations in the framework of a building may prevent doors and windows from opening and closing properly. Similarly, plaster ceilings and walls would crack if the members supporting these surfaces were to deform excessively.

In addition to requiring that a structural material be stiff, it is necessary that deformations resulting from moderate, everyday loads disappear after these loads cease to exist. If this is not the case and deformations are cumulative, repeated applications of even small loads will eventually lead to excessive deformations regardless of whether the material is stiff or not. The property that enables a material to recover its original size and shape after a load is removed and prevents deformations from being cumulative is *elasticity*. All commonly used structural materials, including steel, concrete, and wood, behave elastically for a significant portion of their load-carrying capacity. By comparison, clay is a plastic material; deformations in clay resulting from the application of a load do not disappear when the load is removed.

Although all structural materials, if stressed sufficiently, will eventually fracture, the manner in which the fracture occurs differs from one material to another. Some materials, even though they are stiff at low stresses, experience very large deformations before they actually break, while others fracture suddenly without any indication that the material is approaching its limit of resistance to stress. The materials that remain stiff up to the instant of fracture are known as *brittle* materials. Concrete, stone, cast-iron, and glass are examples of this type of material. By comparison, steel, aluminum, and most metals are *ductile*. These materials, although they are stiff at low stresses, become very flexible and deform considerably before they actually break. In general, a ductile material is preferable to a brittle one because the ductile material gives ample warning of an impending failure, whereas the brittle one does not. Either the excessive deformations in the ductile material are noted visually and the load that is acting is physically removed, or if this is not possible, the structure itself redistributes the stresses from the members that are being overstressed to others that still have additional capacity to resist load. Neither of these alternatives are possible in a brittle material.

In addition to brittleness, there are many other undesirable material characteristics. In fact, almost every structural material has some shortcomings. Steel loses a significant part of its strength at high temperatures and is very susceptible to corrosion. Concrete is brittle and can resist only small amounts of tension, and unreinforced plastics possess neither adequate strength nor adequate stiffness. Fortunately, it is possible by judiciously combining several materials to overcome most of these as well as other material deficiencies. A layer of paint will prevent steel from corroding, and a cover of a few inches of concrete will enable steel to maintain its strength during a fire. A somewhat more sophisticated procedure for enhancing the properties of a material is the combination of concrete and steel to form *reinforced concrete*. The load-carrying capacity of a beam composed solely of concrete is severely limited by the inability of the concrete to resist the tension stresses that occur on the convex side of the beam. As soon as these tension stresses exceed the minimal tensile strength that concrete possesses, the member cracks and collapses. To remedy the situation, it is customary to embed steel bars in the convex side of the beam, where the tension stress occurs. These steel bars then resist the tension stress that exists on the convex side of the beam while the concrete itself resists the compression stress on the concave side. Thus it is possible, by reinforcing a concrete beam with a sufficient amount of steel on its convex side, to overcome the weakness of concrete in tension and produce a beam that can resist tension and compression stresses equally well.

The practice of reinforcing one material with another, used to such great advantage in the production of reinforced concrete, has been applied in other instances as well. In recent years a new group of materials, commonly referred to as *plastics*, have been used extensively in a variety of situations. The attributes of plastics that have led to their widespread use are the ease with which they can be formed into even the most intricate shapes and their high resistance to most forms of corrosion. However, by themselves they are neither strong nor stiff enough to make a satisfactory structural material. They are weak in both tension and compression and deform excessively when subjected to loads. To overcome these shortcomings, it is customary to reinforce the plastics with glass fibers and form a new group of materials called *fiber-reinforced plastics*. By themselves neither glass nor plastics are satisfactory structural materials; the former is too brittle and the latter too weak and too flexible. Together, however, they form a group of composite materials that are relatively strong and stiff and that behave in a ductile manner.

Most material properties that interest the structural engineer can be obtained from a *stress-strain diagram* of the material. For example, let us consider the stress-strain diagram of A36 structural carbon steel shown in Fig. 1.13. The curve indicates that the ultimate strength of the material is approximately 60 ksi. In other words, 60 ksi is the maximum stress that the material can resist without rupturing. Although the material does not fail until the stress reaches 60 ksi, it is undesirable for the stress under ordinary conditions to exceed 36 ksi, the *yield strength* of the material. As long as the stress remains below the yield strength, the material behaves in an elastic manner, i.e., there are no permanent deformations upon removal of the load. Furthermore, by limiting the stress to the region below the yield strength, where the material is very stiff, it is possible to prevent excessive deformations.

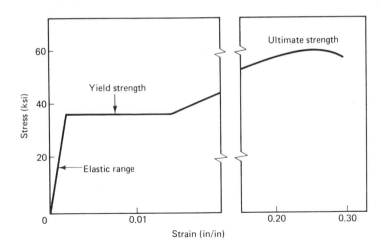

Stress-strain diagram for A-36
carbon steel

FIGURE 1.13

The slope of the stress-strain diagram is a measure of the stiffness of the material. Within the elastic range, which is the range of stress in which the material is expected to be working, it takes a stress of 30 ksi to stretch steel one thousandth of an inch per inch. Steel is thus an extremely stiff material for the range of stresses in which it is expected to perform. Steel is also a very ductile material, as indicated by the fact that it can undergo a total elongation of approximately 25% before it ruptures.

1.4 LOADS

After a decision has been reached regarding the overall size and shape of a structure and the material out of which the structure is to be made, it is necessary to determine the loads that the structure must resist. All loads that can reasonably be expected to act on the structure during its entire life must be considered. This includes the construction phase as well as the time during which the finished structure is in service. From among these loads one must choose those which produce the largest stresses in the structure and then design the individual parts of the structure so that the stresses can be safely resisted.

Loads are usually separated into specific categories. For example, there are static and dynamic loads. A *static load* is one that is applied so slowly that the structure remains at rest during the application of the load. By comparison a *dynamic load* is one that is applied rapidly enough to cause the structure to accelerate and as a consequence gives rise to inertia forces. Thus when dealing with static loads one has only the load itself to consider, whereas a structure subjected to dynamic loads must resist the combined effect of the applied load and the inertia forces.

Another important distinction made among loads is whether they are dead loads or live loads.

Dead loads are those whose magnitude remains constant and whose position does not change with time. They consist of the weight of the structure itself plus the weight of any objects permanently attached to the structure. For example, the dead loads in a building include the weight of the beams and columns that make up the frame, the weight of the walls, floors, and roof, and the weight of permanent fixtures such as plumbing, heating, and air conditioning equipment.

The weights of a few common building materials are given in Table 1.1.

TABLE 1-1 WEIGHTS OF BUILDING MATERIALS

Material	lb/ft^3	kN/m^3
Brick	120	19
Concrete	150	24
Sand	100	16
Steel	490	78
Timber	40	6.4

In comparison with dead loads, *live loads* are those whose magnitude and position do not remain constant. Thus when we consider live loads, we must determine not only the maximum magnitude of the load but also the location of the load that results in the largest stresses. Some examples of live loads are the weight of occupants in a building, the weights of vehicles on a bridge, and the forces exerted by wind on various structures. In each of these instances the position of the force as well as its magnitude are subject to change.

Some commonly encountered live loads are briefly discussed in the following paragraphs.

Building Live Loads

Live loads in buildings are primarily due to the weight of people and movable objects. It is customary to represent these forces by uniformly distributed loads. For example, Table 1.2 lists some loads recommended by the American National Standards Institute (ANSI) Code, to be used in the design of buildings. The intensity of the load is seen to depend on the use to which the area in question will be put.

TABLE 1-2   BUILDING LIVE LOADS

Occupancy	lb/ft^2	kN/m^2
Private apartment	40	1.9
Public room and restaurant	100	4.8
Dance hall	100	4.8
Office	80	3.8
Retail store	100	4.8
Storage warehouse (heavy)	250	12.0
Storage warehouse (light)	125	6.0
Library stacks	150	7.2

Bridge Live Loads

The live loads that act on a highway bridge are mainly due to the weights of the vehicles that cross the bridge. Since bridges are subjected to various combinations of many different vehicles during their lifetime, design codes have developed standard equivalent loadings whose use leads to a safe and economical structure. For example, the specification of the American Association of State Highway and Transportation Officials (AASHTO) recommends that highway bridges be designed to support

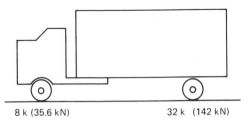

8 k (35.6 kN)   32 k   (142 kN)

H20-44 loading   **FIGURE 1.14**

standard trucks like the one shown in Fig. 1.14. The first number in the truck designation stands for the weight of the vehicle in tons, and the second number specifies the year that the loading was adopted.

Snow Loads

The weight of snow is an important live load in the design of roofs. Two factors that govern the intensity of the snow load to be used are the geographical location of the building and the slope of the roof. Table 1.3 gives snow loads for several areas in the United States. As is to be expected, the snow load is heavier in the northern part of the country than in the south. In addition, the steeper the roof, the smaller the amount of snow that can collect.

TABLE 1-3 SNOW LOAD FOR UNITED STATES (LOAD IN PSF OF HORIZONTAL PROTECTION OF ROOF AREA)

Area	Slope of roof		
	Flat	Medium	High
New England and Great Lakes region	45	30	10
Central states	30	20	10
Southern states	20	15	10

Wind Loads

All exposed surfaces of structures are subject to wind loads. The larger the exposed area, the larger will be the total force to which the structure is subjected. Wind loads are especially important for structures such as highrise buildings, radio towers, and suspension bridges.

The pressure exerted by the wind on a surface depends in general on the velocity of the wind and on the nature of the surface. A relation giving the pressure is

$$p = 0.00256CV^2$$

where p = pressure in psf
C = a factor that depends on the nature of the surface
V = wind velocity in mph

or

$$p = 0.0000473CV^2$$

where p is measured in kN/m² and V in km/hr.

The value of C for buildings with vertical sides has been found to be 1.3. This is a combination of 0.8 due to pressure on the windward side and 0.5 due to suction on the leeward side. Assuming a wind velocity of 100 mph, the wind load on a vertical building would be 33.3 psf.

Other Live Loads

The live loads to which structures can be subjected are almost too numerous to mention. They include forces produced by earthquakes, water pressure on coastal structures, soil pressure on underground structures, and forces that arise when the temperature changes and the structure is not free to expand or contract.

If many structures of the type being designed have already been built, the chances are high that the necessary loads will be given in readily available design codes. However, the loads to which new types of structures are subjected to may not be recorded anywhere and will consequently have to be determined specially for the structure being built.

1.5 HISTORY OF STRUCTURAL ENGINEERING

To see how the various elements of structural engineering that we have considered so far interact with one another and with the needs of society, let us take a brief look at the history of structural engineering.

The Greek and Egyptian temples are among the earliest structures about which we have definite information. These structures were made of stone, and they employed beams and columns as their main structural elements. Unfortunately, stone is relatively weak in tension and is therefore a poor material for making beams. To avoid the formation of cracks on the tension side of the member, it is necessary to make the beam relatively short. The resulting structures (Fig. 1.15) consisted of numerous columns with relatively little useful space between them.

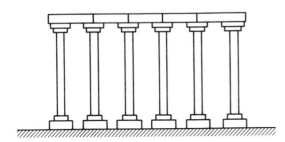
FIGURE 1.15

In comparison with stone beams, stone arches represent a structural engineering achievement of the first order. Although the arch was discovered prior to the Roman era, it was not used extensively until the time of the Roman Empire. The arch is primarily stressed in compression, and stone is an excellent material for resisting compression. The stone arch thus represents an ideal marriage of form and material. Compared to stone beams, whose lengths may not have exceeded 25 feet, stone arches spanned distances of 100 feet and more.

In addition to using the arch to build bridges and aqueducts, the Romans employed it to construct vaulted roofs and domes. The Pantheon, one of the truly magnificent structures built in Roman times, was topped by a dome spanning 142 feet.

Structures spanning distances of this magnitude would not be built again until modern times.

With the decline of the Roman Empire in the fifth century came a decrease in structural engineering activity. The period lasting from about 500 A.D. to 1500 A.D. is known as the Middle Ages or more appropriately the Dark Ages. During this time Western civilization experienced a general decline, and it is thus not surprising that no significant progress was made in the area of structural engineering. The structures that were built continued to employ the stone arch as the major structural form. The most noteworthy among these are the Gothic cathedrals built during the twelfth and thirteenth centuries. These towering structures, soaring toward the heavens, symbolized the spiritual tone of the day.

A most interesting aspect of the cathedrals is the manner in which the roof is supported. Unlike the arches in Roman bridges, which rested on massive abutments, the arched roofs of the cathedrals were supported on tall, thin walls containing glass windows over a sizeable portion of their surface. Since walls such as these are incapable of resisting the outward thrust that the legs of an arch exert on their supports, they were braced by inclined members called *flying buttresses* (Fig. 1.16). The flying buttress was an ingenious device without which many of the splendid cathedrals could not have been constructed.

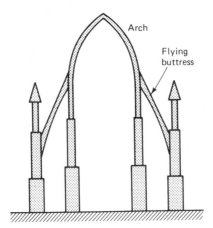

FIGURE 1.16

The fifteenth and sixteenth centuries, known as the Renaissance, were a period of intense activity for Western civilization. It was during these years that many of the foundations of modern society were established. During the Middle Ages man had been more preoccupied with the spiritual than the secular. All this now began to change. Man was no longer considered to be a sinful being whose nature had to be repressed. Instead he felt encouraged to express his individuality and to participate in the molding of his destiny.

As far as structural engineering is concerned, the Renaissance had its greatest impact in the realm of science. It was at this time that Galileo (1564–1642) laid the groundwork for the science of structural analysis. In the first book ever written on the subject, Galileo discussed the concepts of force and moment and analyzed the strength of structural members. Although not all of his work was accurate, he stated correctly that the strength of an axially loaded bar depends on its area and that the strength of a beam is proportional to its depth squared. Another of Galileo's contributions was the influence he exerted on others. Among those following in his footsteps were Robert Hooke (1635–1703), who formulated the law of linear behavior of materials bearing his name; Sir Isaac Newton (1642–1727), who formulated the laws of motion; and Leonhard Euler (1707–1783), who was the first to analyze correctly the buckling of columns.

In addition to the basic work in structural mechanics that was initiated during the Renaissance, an Italian architect, Palladio, is believed to have introduced the use of the truss. The trusses that he and his contemporaries built were modest timber structures whose purpose was more to take the place of simple wooden beams than to replace stone arches. In fact, it was another two hundred years before the use of trusses became widespread. However, the seed of a new and major structural form had been planted.

If the work of men like Galileo and his followers provided the scientific base of the modern era of structural engineering, then the introduction of iron as a building material during the Industrial Revolution can be said to have provided the physical material for modern structural engineering. Man had been using iron for limited purposes ever since the time of the pyramids. However, it was not until the Industrial Revolution occurred, during the latter half of the eighteenth century, that he was able to produce it in sufficient quantity to make its use as a building material feasible.

The first major structure built of iron was a bridge over the Severn River near the English town of Coalbrookdale. Although the cast iron out of which this bridge was constructed is a crude material compared to the steel we use today, a new era in structural engineering had been inaugurated.

The first few iron bridges, like the one at Coalbrookdale, employed the old familiar form of the arch. However, engineers soon recognized the tremendous potential of the new material and started to use it in a variety of novel ways. An especially important use of iron was made in conjunction with the construction of suspension bridges. The early part of the nineteenth century witnessed the erection of several suspension bridges both in England and in America, using iron chains to support the roadway. Among these were Thomas Telford's bridge over the Menai Straits in Wales, Bunel's Clifton Bridge in Bristol, England, and Finley's bridge over the Merrimack River in Newburyport, Massachusetts. Telford's bridge had a span of 579 feet, the longest in the world at the time and probably at least twice as long as any stone arch or timber truss built during the eighteenth century.

In addition to its use in suspension bridges, iron was employed extensively during the nineteenth century for the construction of truss bridges. At the beginning of the eighteenth century most highway bridges were made out of lumber. Many of the covered bridges that are still standing today in parts of the United States date from that period. However, these bridges were neither strong nor stiff enough to serve the needs of the railroads that were starting to develop at the time. As a consequence, by the latter part of the eighteenth century numerous iron truss bridges were being built to replace the old timber structures. The iron truss was a popular structure. It was reasonably strong, easy to build, and not too expensive. Unfortunately many of these structures were poorly designed, and the iron being used at the time was not a very reliable material. As a result, numerous failures occurred. In fact, between 1870 and 1880 as many as 25 bridges failed in the United States during a single year.

The problems plaguing these bridges were eventually eliminated when iron was replaced by steel and structural analysis as we know it today evolved. In 1856, Henry Bessemer, an Englishman, developed the first successful process for producing large quantities of steel economically. Basically, steel is a mixture of iron and a small, carefully controlled amount of carbon. By comparison, the so-called iron used before the introduction of steel contained excessive amounts of carbon, which made it brittle, and various impurities, which made its strength unreliable.

In addition to the introduction of steel as a structural material, the latter part of the nineteenth century saw tremendous progress in the science of structural analysis. Men such as Mohr and Muller-Breslau in Germany, Maxwell in England, Castigliano in Italy, and Greene in America developed routine procedures for analyzing both determinate and indeterminate structures. Together with the general availability of steel, these advances in structural analysis paved the way for structural engineering in the twentieth century. The long-span suspension bridges and the highrise buildings, two of the most significant achievements of twentieth-century structural engineers, would not have been possible without the availability of a reliable high-strength material such as steel and reliable methods of structural analysis.

2

Calculation of Reactions

A bridge in Oregon.
(*Courtesy of Oregon
Department of
Transportation.*)

2.1 INTRODUCTION

The first step in the analysis of a structure usually consists of drawing a *free-body diagram*. This is a simplified picture of the structure, isolated from its supports, on which are shown all the external forces that act on the structure. In general, these forces include the applied loads and the reactions that the supports exert on the structure. For example, Fig. 2.1b depicts the free-body diagram of the beam in Fig. 2.1a. Shown acting on the free-body diagram are the two known loads and the three unknown reactions that the supports apply to the beam. Since the beam is in equilibrium under the action of these forces, one can use the conditions of equilibrium to solve for the unknown reactions in terms of the known loads. A review of the conditions of equilibrium as well as the manner in which they are used to determine unknown reactions is the main subject of this chapter.

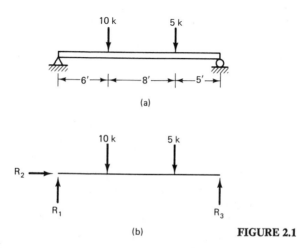

(a)

(b)   FIGURE 2.1

2.2 EQUATIONS OF EQUILIBRIUM

In this book we will primarily be concerned with structures at rest. For such a structure, the sum of the forces in any direction must vanish and the moment about any axis of all the forces acting on the structure must be zero. In statics it was demonstrated that these conditions of equilibrium are satisfied if the sum of the forces along any three mutually perpendicular axes are zero and if the moments of the forces about these three axes are also zero. If we denote the axes by x, y, and z, the above requirements can be expressed analytically by the equations

$$\sum F_x = 0, \qquad \sum F_y = 0, \qquad \sum F_z = 0$$
$$\sum M_x = 0, \qquad \sum M_y = 0, \qquad \sum M_z = 0 \tag{2.1}$$

Equations (2.1) apply to a three-dimensional structure. However, in many instances structures can be represented by a planar system. In other words, the

20

structure as well as the forces acting on it can be assumed to lie in a plane. If this plane is the x-y plane, then equilibrium is established if

$$\Sigma F_x = 0, \qquad \Sigma F_y = 0, \qquad \Sigma M_0 = 0 \tag{2.2}$$

Strictly speaking, one should sum moments about the z-axis. However, when dealing with a planar system, one usually sums moments about the intersection of the z-axis and the plane in which the structure lies. In other words, one simply writes a moment equation about a point-0.

Sometimes all the forces acting on a structure pass through a common point. Such a force system is referred to as a concurrent set of forces, and equilibrium is established if

$$\Sigma F_x = 0, \qquad \Sigma F_y = 0, \qquad \Sigma F_z = 0 \tag{2.3}$$

for a three-dimensional structure, and if

$$\Sigma F_x = 0, \qquad \Sigma F_y = 0$$

for a two-dimensional structure.

2.3 TYPES OF SUPPORTS AND RESTRAINTS

In constructing the free-body diagram of a structure, it is necessary to replace the supports of the structure by the forces that they exert on the structure. The most common types of supports encountered are the roller, the hinge, the fully fixed support, and the cable. Each of these supports is illustrated in Table 2.1.

TABLE 2.1

Types of supports		
Roller		
Hinged support		
Fixed support		
Cable		

1. At a *roller*, a member is free to translate parallel to the surface on which the roller rests and free to rotate. Motion is restrained only in the direction normal to the surface that supports the roller. A roller is thus only capable of applying a force normal to the surface on which it rests.

2. A *hinged support* allows the member to rotate freely but prevents any type of translation. It restrains the member by applying forces both normal and parallel to the supporting surface, i.e., a horizontal and a vertical force.

3. A *fixed support* allows for no motion at all. The member can neither rotate nor translate. This type of support is able to exert a moment as well as a horizontal and vertical force on a member.

4. A *cable* can prevent motion only along its axis and then only if the motion tends to put the cable in tension. In other words, a cable can exert a pull but not a push.

2.4 STABILITY AND DETERMINACY

A structure for which all the unknown reactions can be determined using the equations of equilibrium is referred to as a *determinate structure*. For example, the structure in Fig. 2.2a possesses three unknown reactions, and since it is a plane structure, there are available three equations of equilibrium to solve for these reactions. Thus the structure in Fig. 2.2a is said to be determinate. If the roller at *C* is replaced by a hinge, as indicated in Fig. 2.2b, the number of unknown reactions is increased from three to four. However, there are still only three independent equations of equilibrium that can be used to evaluate the reactions. A structure of this type, which possesses more unknown reactions than equations of equilibrium, is referred to as an *indeterminate structure*. It is still possible to determine the reactions in such a structure. However, equations other than those dealing with equilibrium must be employed in carrying out the solution. We will consider this type of structure in the latter part of the book.

A third possibility is that the pin at A is replaced by a roller as shown in Fig. 2.2c. In this case there is nothing to prevent the structure from moving toward the right when subjected to the applied loads. A structure such as this, in which there are an insufficient number of reactions to prevent motion from taking place, is called an *unstable structure*.

In general, a plane structure is both stable and determinate if it is supported by three reaction components that are neither parallel nor concurrent. Two reaction components may be able to prevent motion of a structure for a specific loading but they are usually unable to do this for all loadings. For example, the structure in Fig. 2.3a is stable as long as it is only subjected to a vertical load. However, the structure is unable to resist a horizontal load and it is therefore said to be unstable. In other words, a structure is classified as being stable only if it can resist any and all possible loads without moving. Although at least three reaction components are necessary to provide stability this number is not always sufficient. The structure in Fig. 2.3b has three reactions. Nevertheless, because the reactions are all vertical the structure cannot

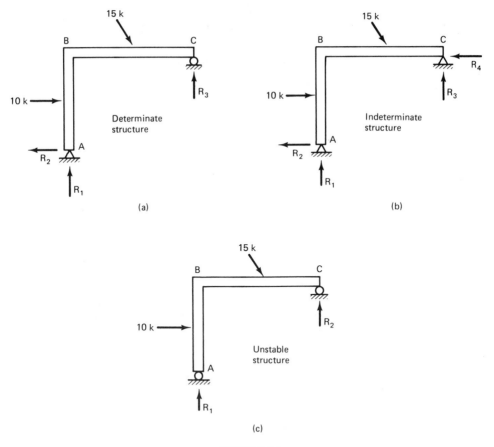

FIGURE 2.2

resist a horizontal load and is therefore unstable. Similarly, the structure in Fig. 2.3c has three reactions, yet it is unstable. The lines of action of the three reactions all pass through a common point making it impossible for them to resist an applied moment about this point.

The structure in Fig. 2.4a is determinate because it has three unknown reactions and there are three equations of equilibrium available to determine the reactions. By comparison the structure in Fig. 2.4b is indeterminate. It has four unknown reactions, but only three equations of equilibrium are applicable. It may appear from these two examples that a plane structure possessing more than three reactions is always indeterminate. However, this is not the case. Sometimes, so-called compound structures, formed by connecting in a nonrigid manner two or more simple structures can have more than three reactions and still be determinate. For example, the structure in Fig. 2.5 has four reactions but it is nevertheless determinate. One has available for determining the reactions, in addition to the three equations of equilibrium for the structure as a whole, an equation of construction due to the hinge at A. Since the hinge cannot transmit a moment, the moment of all the forces either to the right or to the left of the hinge must be zero about the hinge.

(a)

(b)

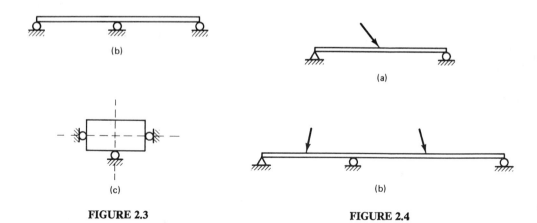

(c)

FIGURE 2.3

(a)

(b)

FIGURE 2.4

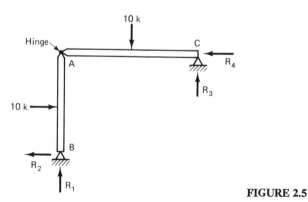

FIGURE 2.5

To get a clear picture of the effect that the hinge at A has on the structure in Fig. 2.5, let us compare the free bodies into which the structure has been subdivided shown in Fig. 2.6a with those in Fig. 2.6b. The former correspond to the structure with a hinge at A and the latter to the same structure without the hinge at A. With a hinge at A, there exist a total of six unknown forces among the two free bodies. Since we can write three equations for each free body we have available a total of six equations to solve for the six unknowns, i.e. the structure is determinate. By comparison, if there is no hinge at A we have seven unknown forces between the two free bodies or one more unknown than we have equations of equilibrium. The structure without the hinge at A is thus indeterminate.

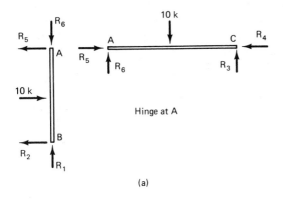

Hinge at A

(a)

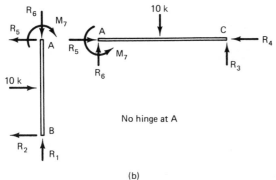

No hinge at A

(b)

FIGURE 2.6

2.5 CALCULATION OF REACTIONS—SIMPLE STRUCTURES

Once the free-body diagram of a structure has been drawn, the calculation of the reactions is usually the next step. Since most of the analysis that follows the calculation of the reactions depends on the reactions, the importance of this initial step cannot be overemphasized

Example 2.1

As an illustration of the procedure followed in determining reactions, let us consider the structure in Fig. 2.7a. The structure is acted on by two known loads and supported by a hinge at A and a roller at B. In Fig. 2.7b, the supports have been replaced by a horizontal and a vertical reaction at A and a vertical reaction at B. In addition the 60 k load has been resolved into its vertical and horizontal components. For a plane structure such as the one being considered, there are available three independent equations of equilibrium: $\sum F_x = 0$, $\sum F_y = 0$ and $\sum M_0 = 0$. Since the number of reactions, three in this case, does not exceed the number of available equations of equilibrium, the structure is determinate and the reactions can be evaluated using the equations of equilibrium.

Although the calculation of the reactions can be carried out in a haphazard

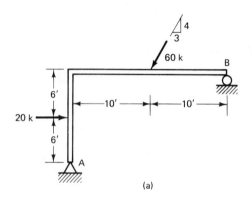

(a)

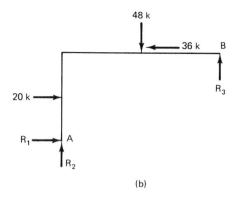

(b)

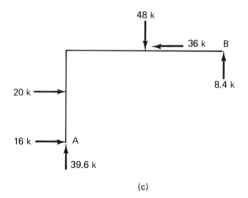

(c)　　　　　　　　　　　　　**FIGURE 2.7**

manner, it is possible by applying the equations of equilibrium judiciously to keep the numerical work to a minimum. For example, the equation of horizontal equilibrium and a moment equation about point A each contain only one unknown reaction. Using these equations, it is therefore possible to immediately solve for R_1 and R_3. Thus

$$\Sigma F_x = 0$$
$$R_1 = 36 - 20 = 16$$

$$R_1 = 16 \text{ k} \rightarrow$$

$$\sum M_A = 0$$
$$-R_3(20) + 48(10) - 36(12) + 20(6) = 0$$
$$R_3 = 8.4$$

$$R_3 = 8.4 \text{ k} \uparrow$$

Once R_3 is known, the equation of vertical equilibrium can be used to calculate R_2:

$$\sum F_y = 0$$
$$R_2 = 48 - 8.4 = 39.6$$

$$R_2 = 39.6 \text{ k} \uparrow$$

Although only three independent equations can be written to determine unknown reactions for a plane structure, one has a choice regarding which equations to use. For instance, it is usually possible to substitute moment equations for horizontal and vertical equilibrium equations. Thus we could have solved for R_2 by writing a moment equation about point B instead of using the equation of vertical equilibrium.

Because of the importance of obtaining the correct values for the reactions, one should always take the time required to check these calculations. The check is performed using an equation not employed in the calculation of the reactions. Thus we will check the values obtained for the reactions by writing a moment equation about point B.

$$\sum M_B = 0$$
$$39.6(20) - 16(12) - 20(6) - 48(10) = 0$$
$$0 = 0$$

O.K.

To avoid confusion, it is helpful to show all of the reactions on a free-body diagram of the structure, after the calculations have been completed. This is done in Fig. 2.7c.

2.6 CALCULATION OF REACTIONS—
COMPOUND STRUCTURES

Whereas some structures, like the one in Fig. 2.7, consist of a single rigid member, others are made up of a number of members connected to one another in a nonrigid manner. For example, the system depicted in Fig. 2.8a consists of two bars, AC and BD, connected to each other by a pin at B. At first sight this structure, having four unknown reactions, would appear to be indeterminate. However, the pin connection at B provides a condition of connection which when combined with the three basic equations of equilibrium for a plane structure makes it possible to determine all four reactions.

Example 2.2

To illustrate the procedure used to calculate reactions for a compound structure, let us determine the reactions for the structure in Fig. 2.8a. A very direct solution to the problem would be simply to write equations of vertical, horizontal, and moment equilibrium for the entire structure in addition to summing moments about the pin at

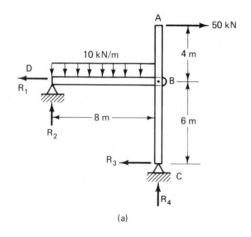

(a)

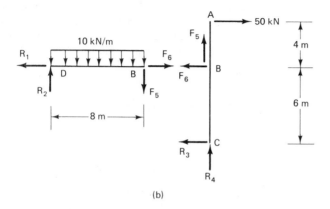

(b)

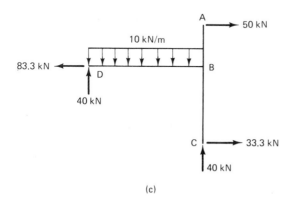

(c)

FIGURE 2.8

B of the forces either to the left or to the right of *B*. The disadvantage of this approach is that it entails the solution of simultaneous equations, which is undesirable if the calculations are being carried out without the aid of a computer. We will therefore approach the solution in a somewhat different manner. To begin, let us separate the structure into its basic components as indicated in Fig. 2.8b. We can now write equations of equilibrium for either of the two free bodies in Fig. 2.8b or for the entire structure in Fig. 2.8a. It should be noted that the free bodies of the individual bars have acting on them, in addition to the applied loads and reactions at C and D, the two components of the force that they apply to one another at B. In accordance with the principle of action and reaction, the forces at B acting on bar BD are equal and opposite to those acting on bar AC.

Starting with the free body of bar ABC, we can sum moments about B and solve for R_3. Thus

$$\Sigma M_B = 0$$
$$R_3(6) + 50(4) = 0$$
$$R_3 = -33.3$$
$$R_3 = 33.3 \text{ kN} \rightarrow$$

The negative sign in front of the answer indicates that the assumed direction of R_3 is incorrect.

At this point we can proceed in one of two ways. We can either turn to the free body of the entire structure and solve for R_4 by summing moments about D and R_1 and R_2 by writing equations of horizontal and vertical equilibrium, or we can continue using the free bodies of the individual bars. Using the former approach, we turn to the free body of the entire structure.

$$\Sigma M_D = 0$$
$$10(8)(4) + 50(4) + (-33.3)(6) - R_4(8) = 0$$
$$R_4 = 40$$
$$R_4 = 40 \text{ kN} \uparrow$$

$$\Sigma F_x = 0$$
$$50 - (-33.3) - R_1 = 0$$
$$R_1 = 83.3$$
$$R_1 = 83.3 \text{ kN} \leftarrow$$

$$\Sigma F_y = 0$$
$$R_2 + 40 - 10(8) = 0$$
$$R_2 = 40$$
$$R_2 = 40 \text{ kN} \uparrow$$

We will now check the values obtained for the four reactions by summing moments about point A for the entire structure.

$$\Sigma M_A = 0$$
$$(-33.3)(10) - 10(8)(4) + 40(8) + 83.3(4) = 0$$
$$0 = 0$$

O.K.

A free-body diagram of the entire structure, including all the reactions, is shown in Fig. 2.8c.

Example 2.3

Determine the reactions for the structure in Fig. 2.9.

The structure in Fig. 2.9a is known as a *cantilever structure*. The presence of two hinges in the center span makes the structure determinate. This type of structure appealed to nineteenth-century bridge builders, who had not yet mastered the science of analyzing indeterminate structures.

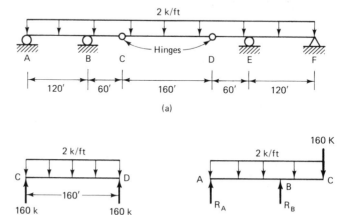

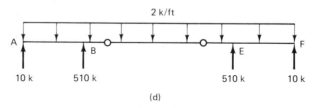

FIGURE 2.9

It is best to begin the analysis of the structure by considering a free body of the section between the hinges, as shown in Fig. 2.9b. Since hinges cannot transmit moments, only vertical forces are applied to member CD by the adjacent segments of the structure. Because of symmetry, each of these forces is equal to 160 k.

We are now ready to consider the free body of span ABC, shown in Fig. 2.9c. This member has acting on it, in addition to the applied distributed load and the reactions at A and B, a 160 k force at C. In accordance with the principle of action and reaction, if ABC pushes up on CD with a force of 160 k, then CD must push down on ABC with a force of 160 k. Taking moments about point A, we obtain

$$360(90) + 160(180) - R_B(120) = 0$$

$$R_B = 510 \uparrow$$

and from vertical equilibrium

$$R_A + 510 - 360 - 160 = 0$$

$$R_A = 10 \uparrow$$

Since the entire structure is symmetric about its centerline,

$$R_E = R_B = 510 \uparrow, \qquad R_F = R_A = 10 \uparrow$$

A free body of the entire structure, showing all reactions, appears in Fig. 2.9d.

PROBLEMS

2.1 to 2.19. Determine the reactions.

2.1.

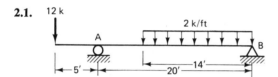

2.2.

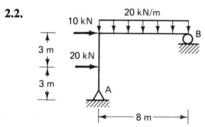

2.3.

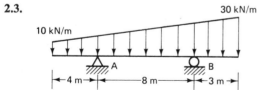

2.4.

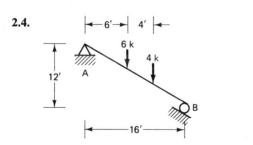

2.5.

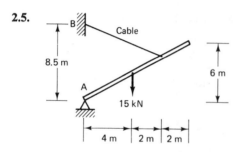

2.6.

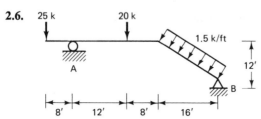

2.7.

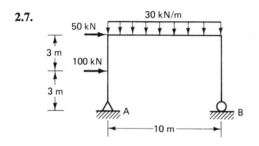

2.8.

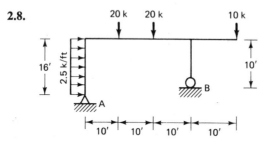

2.9.

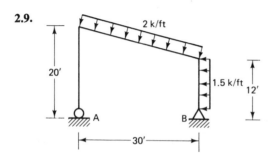

2.10.

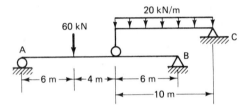

2.11.

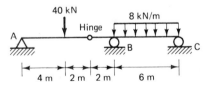

2.12.

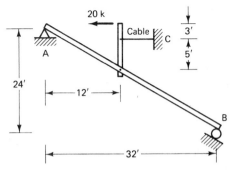

2.13.

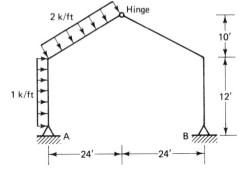

2.14.

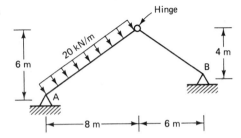

2.15.

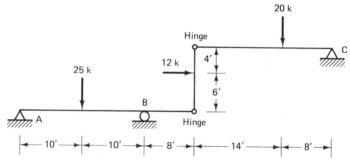

2.16.

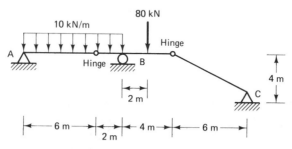

2.17.

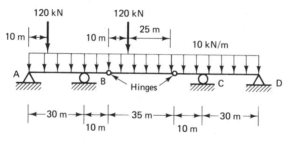

2.18.

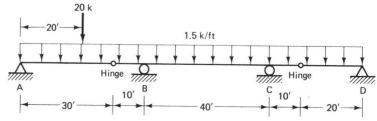

2.19.

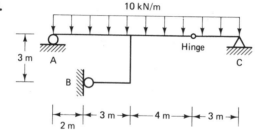

3

Plane Trusses

Truss Bridge, Sewickley, Pa. (*Courtesy of U.S. Steel Corporation.*)

3.1 INTRODUCTION

A truss is a system of relatively slender members, arranged in the form of one or more triangles, which transfers loads by developing axial forces in its members. Trusses are commonly employed in bridges, towers, and roof structures. They use material very efficiently and are consequently economical for spanning distances up to several hundred feet.

3.2 SIMPLIFYING ASSUMPTIONS

Our primary objective when analyzing a truss is to determine the forces developed in the individual members by a set of externally applied loads. An analysis of this sort can be greatly simplified without unduly impairing the accuracy of the results by making the following assumptions:

1. The members are connected to each other at their ends by frictionless pins, i.e., only a force and no moment can be transferred from one member to another.
2. External loads are applied to the truss only at its joints.
3. The centroidal axes of the members meeting at a joint all intersect at a common point, namely, the point where the members are assumed to be pinned to one another.

It is a direct consequence of these idealizations that the members of a truss can be assumed to be subjected to axial forces only. Shown in Fig. 3.1a is the free-body diagram of a single member of a truss. Since such a member is assumed to be connected, at its ends, to other members by pins, only forces and no moments can be applied at these ends. In addition, since all external loads are assumed to be applied to the truss at its joints, no loads can be acting on the member between its end points. If we resolve the forces acting on the ends of the member into components along the

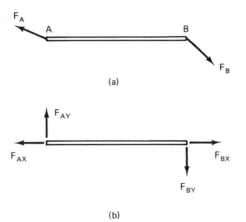

(a)

(b)

FIGURE 3.1

x and y axes (as shown in Fig. 3.1b) and apply the equations of equilibrium $\sum F_x = 0$, $\sum F_y = 0$, and $\sum M_0 = 0$, it becomes apparent that equilibrium can be achieved only if the components in the y-direction are zero and if the components in the x-direction are equal and opposite to each other. Furthermore, since the points of application of the forces acting on the ends of the member are assumed to lie along the centroidal axis of the member, the x-components of these forces will coincide with the centroidal axis and the member will be subject to axial tension or compression only.

The assumption regarding the manner in which the external loads are applied to a truss is generally valid. In a roof system (Fig. 3.2) the roofing material is usually supported by members running transverse to the trusses, called *purlins*. These purlins tie into the trusses at their joints, making it possible for the load to be taken from the

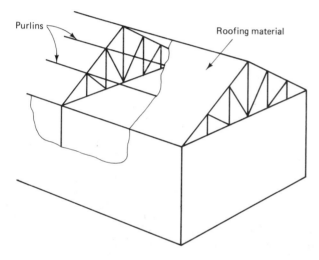

Purlins

Roofing material

FIGURE 3.2

roof and applied to the trusses at their joints only. In a similar manner, the deck of a bridge is supported by lateral members called *floor beams*, which transmit loads from the deck of the bridge to the trusses at their joints (Fig. 3.3). Of all the loads that either a roof or a bridge truss must resist, the weight of the individual members is probably the most significant load not applied directly at the joints. However, the weight of a truss is usually small compared to the other applied loads, and the error introduced by either neglecting it entirely or assuming it to be applied at the joints is fairly small.

In most modern trusses, the members are not connected to each other by means of frictionless pins. Instead, truss members are usually riveted, bolted, or welded to one another at the joints. Therefore there is a possibility of actual truss members being subjected to bending in addition to axial loading. However, since the members of a truss are relatively slender and consequently not very stiff in bending, and since trusses are designed so that the centroidal axes of the members meeting at any joint do go through a common point, the members of a real truss will primarily be subjected to axial forces. The stresses due to these axial forces are referred to as *primary stresses*,

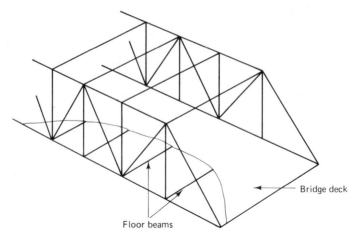

FIGURE 3.3

and under ordinary circumstances it suffices to base the design of a truss on these axial stresses. The relatively small bending stresses that may occur in members of actual trusses, as a result of the fact that they differ from the idealized structures we have pictured here, are referred to as *secondary stresses*. Although these may have to be considered in some instances, we will ignore them in this chapter.

3.3 BASIC CONCEPTS

The object of analyzing a truss is to determine the forces produced in its members by a set of external loads. Since these forces are internal when one considers the truss as a whole, it is necessary in the course of the analysis to take free bodies of portions of the truss. Such a free body, of a portion of a truss, will be acted on by the bar forces corresponding to the members that have been cut in the process of creating the free body. If one then writes equations of equilibrium for the free body, one can use these equations to evaluate the bar forces.

There are two well-known methods of passing sections through a truss in order to produce useful free bodies of portions of the truss. In one, known as the *method of joints*, we consider free bodies of individual joints; in the other, known as the *method of sections*, we pass a section that bisects the truss and consider the free body on either side of the section. It is usually advantageous to employ the method of sections when one wishes to determine the forces in only a few isolated members and to use the method of joints when it is desired to calculate all the bar forces in a truss.

Although we are interested in determining the bar forces themselves, it is necessary to deal with their horizontal and vertical components during the analysis. To see how the components of a bar force are obtained, let us consider the force in member AB of the truss depicted in Fig. 3.4. The horizontal component of the force is given by

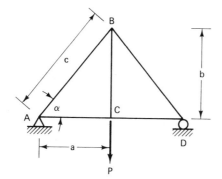
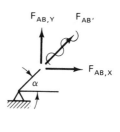

FIGURE 3.4

$$F_{AB,x} = F_{AB} \cos \alpha = F_{AB}\left(\frac{a}{c}\right)$$

and the vertical component by

$$F_{AB,y} = F_{AB} \sin \alpha = F_{AB}\left(\frac{b}{c}\right)$$

The horizontal or vertical component of a bar force can thus be obtained by multiplying the magnitude of the bar force by the ratio of the horizontal or vertical projection of the member to its entire length. As a rule it is easier to use these ratios than to calculate the angles such as α and their sines and cosines.

Since the force in a bar can be such as to produce either tension or compression in the member, it is necessary to have a sign convention for designating which of these two conditions exists. We will do this by using (T) to denote tension and (C) to denote compression. An alternative convention is to denote tension forces by a plus sign (+) and compression forces by a minus sign (−).

3.4 METHOD OF JOINTS

In the method of joints, a section is passed completely around a joint, cutting all the bars meeting at the joint and isolating the joint from the rest of the truss, as shown in Fig. 3.5a. The free body of the joint produced in this manner will have acting on it a set of concurrent forces consisting of bar forces and externally applied loads or reactions. For example, the free body of joint A, shown in Fig. 3.5a, has acting on it the left-hand reaction of the truss and bar forces F_{AB} and F_{AC}. If one considers free bodies of the bars as well as the joints (Fig. 3.5b), it becomes evident that members in compression such as bar AB push on the joints at their extremities and that tension members such as bar AC pull on the joints at their extremities.

Since joint A is in equilibrium, the equations $\sum F_x = 0$ and $\sum F_y = 0$ can be used to determine the two unknown forces acting on the joint. Of course, not all joints in a truss involve only two member forces. However, for most trusses one can

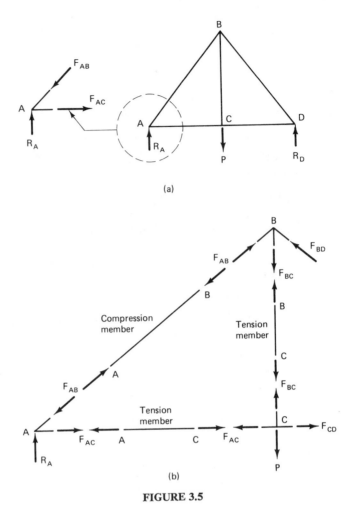

(a)

(b)

FIGURE 3.5

proceed from one joint to another in such a manner that there are never more than two unknown bar forces acting on the joint being investigated. To see how the method of joints is carried out, let us consider Example 3.1.

Example 3.1

It is required to calculate the bar forces for the truss in Fig. 3.6a using the method of joints.

The first step, in this as in any other analysis, is to determine the reactions. Since an error in the reactions renders the remainder of the analysis useless, one cannot overemphasize the importance of this step. To minimize the chances of error, the calculation of the reactions should always be checked. We will obtain the reactions by writing two moment equations and then check the results using an equation of vertical equilibrium. Thus

40

$$\Sigma \, M_A = 0$$
$$-R_D(60) + 24(40) + 12(20) = 0$$
$$R_D = 20$$

$$R_D = 20 \text{ k} \uparrow$$

$$\Sigma \, M_D = 0$$
$$R_A(60) - 12(40) - 24(20) = 0$$
$$R_A = 16$$

$$R_A = 16 \text{ k} \uparrow$$

$$\Sigma \, F_y = 0$$
$$16 + 20 = 12 + 24$$
$$0 = 0$$

O.K.

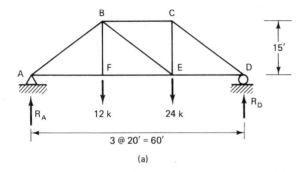

(a)

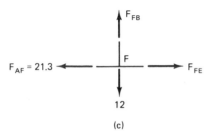

(b)

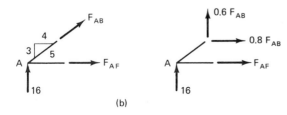

(c)

FIGURE 3.6

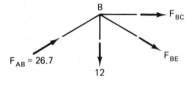
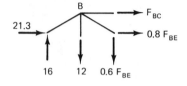

(d)

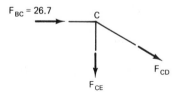
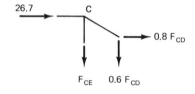

(e)

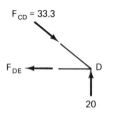
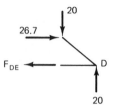

(f)

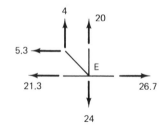

(g)

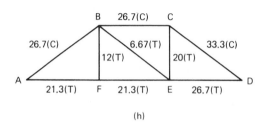

(h)

FIGURE 3.6 (Continued.)

Having determined the reactions, we are now ready to proceed with the calculation of the bar forces. Since we will need the slopes of the members in carrying out these calculations, they have been determined and noted in Fig. 3.6a. We begin by considering joint A, a joint on which there are only two unknown bar forces acting. Fig. 3.6b depicts a free body of joint A. Acting on the free body are the left-hand reaction of 16 k and bar forces F_{AB} and F_{AF}. It is customary to assume all unknown bar forces to be tension forces. Accordingly the forces in the figure are shown pulling on the joint. Knowing the slope of member AB, one can replace F_{AB} by its vertical and horizontal components as indicated.

When using equations of equilibrium to solve for unknown bar forces at a joint it is desirable, whenever possible, to write the equations in such a way that each equation leads to a solution for an unknown without having to consider the two equations simultaneously. For example, by writing the equation of vertical equilibrium first, we can immediately solve for F_{AB}. Thus

$$\Sigma F_y = 0$$
$$16 + 0.6F_{AB} = 0$$
$$F_{AB} = -26.7$$
$$F_{AB} = 26.7 \text{ k (C)}$$

The negative sign of the answer indicates that F_{AB} was assumed to act in the wrong direction. In other words, member AB actually pushes on the joint, which means that the member is in compression.

Having determined F_{AB}, we can now use the equation of horizontal equilibrium to calculate F_{AF}.

$$\Sigma F_x = 0$$
$$F_{AF} + 0.8F_{AB} = 0$$
$$F_{AF} = -0.8(-26.7) = 21.3$$
$$F_{AF} = 21.3 \text{ k (T)}$$

This time the answer is positive, indicating that member AF pulls on the joint as assumed and that the member is therefore in tension.

At this stage in the analysis, joint B has three unknown forces acting on it, whereas joint F has only two unknown forces acting on it. We will therefore consider F next. As shown in Fig. 3.6c, this joint has acting on it a 12 k load, the two unknown bar forces F_{FB} and F_{FE} and the previously determined force $F_{AF} = 21.3$ k. The unknown bar forces are readily determined using equations of vertical and horizontal equilibrium. Thus,

$$\Sigma F_y = 0$$
$$F_{FB} - 12 = 0$$
$$F_{FB} = 12$$
$$F_{FB} = 12 \text{ k (T)}$$

$$\Sigma F_x = 0$$
$$F_{FE} - 21.3 = 0$$
$$F_{FE} = 21.3$$
$$F_{FE} = 21.3 \text{ k (T)}$$

Since of the four bar forces acting on joint B only two are still unknown, we can now proceed with the analysis of joint B. The free body of this joint is shown in Fig. 3.6d. It should be noted that whereas the unknown forces are assumed to be tension, the known forces in bars AB and FB are shown acting in their correct directions. As before, we replace all forces by their horizontal and vertical components and write equations of horizontal and vertical equilibrium.

$$\Sigma F_y = 0$$
$$16 - 12 - 0.6F_{BE} = 0$$
$$F_{BE} = 6.67$$

$$F_{BE} = 6.67 \text{ k (T)}$$

$$\Sigma F_x = 0$$
$$21.3 + 0.8(6.67) + F_{BC} = 0$$
$$F_{BC} = -26.7$$

$$F_{BC} = 26.7 \text{ k (C)}$$

Next we consider the free body of joint C, shown in Fig. 3.6e, and determine F_{CE} and F_{CD}.

$$\Sigma F_x = 0$$
$$0.8F_{CD} + 26.7 = 0$$
$$F_{CD} = -33.3$$

$$F_{CD} = 33.3 \text{ k (C)}$$

$$\Sigma F_y = 0$$
$$F_{CE} + 0.6(-33.3) = 0$$
$$F_{CE} = 20.0$$

$$F_{CE} = 20.0 \text{ k (T)}$$

Finally, the value of F_{DE} is obtained using the free body of joint D, shown in Fig. 3.6f.

$$\Sigma F_x = 0$$
$$26.7 - F_{DE} = 0$$
$$F_{DE} = 26.7$$

$$F_{DE} = 26.7 \text{ k (T)}$$

Having determined all the bar forces, we should now check our results. In obtaining the bar forces we did not make use of the equation of vertical equilibrium at joint D and the equations of vertical and horizontal equilibrium at joint E. These are not independent equations. However, they can be used to check our calculations. The reason we did not make use of these three equations is that the reactions which could have been determined using joint equations were instead determined using a free body of the entire truss.

For joint D, shown in Fig. 3.6f,

$$\Sigma F_y = 0$$
$$20 - 20 = 0$$

and for joint E, shown in Fig. 3.6g,

$$\sum F_y = 0$$
$$4 + 20 - 24 = 0$$
$$\sum F_x = 0$$
$$26.7 - 21.3 - 5.3 = 0.1 \qquad\qquad \text{O.K.}$$

When the analysis of a truss has been completed, it is useful to record all the bar forces on a diagram of the truss, as shown in Fig. 3.6h.

3.5 METHOD OF SECTIONS

In the method of sections, a section is passed completely through the truss, dividing it into two free bodies. Each of these free bodies will have acting on it bar forces at the cut members and externally applied loads. By writing equations of equilibrium for the free body on either side of the section, it is possible to solve for the forces in the members cut by the section. Unlike the method of joints, the method of sections allows one to determine the force in a member located anywhere in the truss without first calculating the forces in other members. To see how the method of sections is applied, let us consider the following illustrative examples.

Example 3.2

Using the method of sections, it is required to determine the forces in bars BC, HG, and BG for the truss in Fig. 3.7a.

The first step is to calculate the reactions at A and E. Next a section is passed through the truss cutting members BC, BG, and HG, whose forces we wish to determine. In order to calculate these forces, we will consider the free body to the left of the section, shown in Fig. 3.7b. As before, the unknown bar forces are initially assumed to be tension forces. If we write a moment equation about G, a point through which the lines of action of two of the unknown forces go, we can immediately solve for F_{BC}. Thus

$$\sum M_G = 0$$
$$6F_{BC} + 60(16) - 40(8) = 0$$
$$F_{BC} = -106.7$$
$$F_{BC} = 106.7 \text{ kN (C)}$$

In a similar manner we can solve directly for F_{HG} by summing moments about point B.

$$\sum M_B = 0$$
$$60(8) - 6F_{HG} = 0$$
$$F_{HG} = 80$$
$$F_{HG} = 80 \text{ kN (T)}$$

Finally, the force in bar BG is obtained by writing an equation of vertical equilibrium. Thus

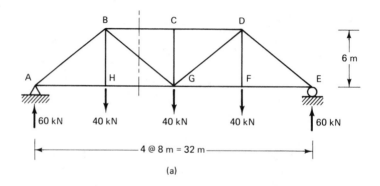

(a)

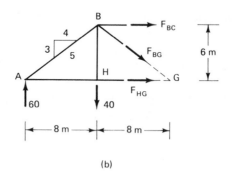

(b)

FIGURE 3.7

$$\Sigma \, F_y = 0$$
$$60 - 40 - 0.6F_{BG} = 0$$
$$F_{BG} = 33.3$$

$$F_{BG} = 33.3 \text{ kN (T)}$$

To check our results, we write an equation of horizontal equilibrium.

$$\Sigma \, F_x = 0$$
$$F_{BC} + F_{HG} + 0.8F_{BG} \quad = 0$$
$$-106.7 + 80 + 0.8(33.3) = 0$$
$$-0.1 = 0$$

Example 3.3

Using the method of sections, determine the forces in bars BC, BG, and HG for the truss in Fig. 3.8a.

As in the previous example, the reactions are calculated first. A section is then passed through the truss as indicated, resulting in the free body shown in Fig. 3.8b. To determine the force in member HG, we sum moments about point B. Thus

$$\Sigma \, M_B = 0$$
$$30(20) - F_{HG}(10) = 0$$
$$F_{HG} = 60$$

$$F_{HG} = 60 \text{ k (T)}$$

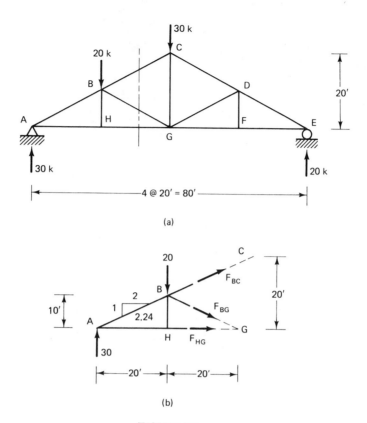

FIGURE 3.8

In a similar manner we can calculate F_{BC} by taking moments about point G. This requires taking moments of force F_{BC} about G. Although it would be possible for us to obtain this moment by determining the perpendicular distance from G to F_{BC}, it is easier to resolve F_{BC} into vertical and horizontal components at either B or C and then, knowing the distances from B and C to G, calculate the moments of these components about G. In fact, only the horizontal component of F_{BC} has a nonzero moment about G, if we resolve F_{BC} into components at point C. Thus

$$\sum M_G = 0$$
$$30(40) - 20(20) + 0.89F_{BC}(20) = 0$$
$$F_{BC} = -44.9$$

$$F_{BC} = 44.9 \text{ k (C)}$$

To determine F_{BG}, we will sum moments about point A. The moment of F_{BG} is obtained by resolving the force into components at point G.

$$\sum M_A = 0$$
$$20(20) + 0.45F_{BG}(40) = 0$$
$$F_{BG} = -22.2$$

$$F_{BG} = 22.2 \text{ k (C)}$$

47

To check our results, we write an equation of horizontal equilibrium.

$$\Sigma F_x = 0$$

$$F_{HG} + 0.89F_{BG} + 0.89F_{BC} = 0$$

$$60 + 0.89(-22.2) + 0.89(-44.9) = 0$$

$$0.2 = 0$$

3.6 METHODS OF JOINTS AND SECTIONS COMBINED

Even when one wishes to determine all the bar forces in a truss, it may be desirable to use both the method of joints and the method of sections. In the following example we will employ the method of joints to determine some of the bar forces and the method of sections to calculate others.

Example 3.4

Determine all the bar forces for the truss in Fig. 3.9a.

Let us begin by calculating F_{AF} and F_{AB} using the free body of joint A in Fig. 3.9b. Since both forces have horizontal as well as vertical components, it is not possible to avoid solving simultaneous equations.

$$\Sigma F_x = 0$$

$$0.447F_{AB} + 0.894F_{AF} - 30 = 0$$

$$\Sigma F_Y = 0$$

$$0.894F_{AB} + 0.447F_{AF} + 25 = 0$$

Solving these equations gives

$$F_{AB} = -59.7 \text{ kN}, \qquad F_{AF} = 63.4 \text{ kN}$$

Next, F_{BC} is determined using the free body in Fig. 3.9c.

$$\Sigma M_F = 0$$

$$0.894F_{BC}(9) + 25(12) + 30(6) + 30(6) - 20(6) = 0$$

$$F_{BC} = -67.1 \text{ kN}$$

Using the free body of joint C in Fig. 3.9d gives

$$F_{CF} = 20.0 \text{ kN}, \qquad F_{CD} = -67.1 \text{ kN}$$

and from the free body of joint B in Fig. 3.9e we obtain

$$\Sigma F_y = 0$$

$$0.894(59.7) - 20 - 0.447(67.1) - 0.707F_{BF} = 0$$

$$F_{BF} = 4.7 \text{ kN}$$

To determine F_{EF}, we make use of the free body in Fig. 3.9f.

$$\Sigma M_D = 0$$

$$0.894F_{EF}(6) + 0.447F_{EF}(6) - 55.0(6) = 0$$

$$F_{EF} = 41.0$$

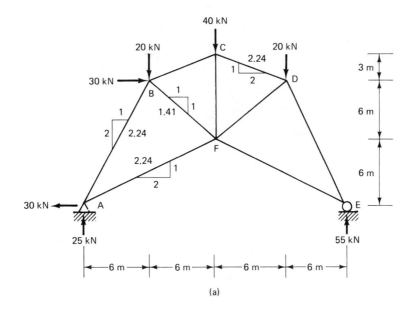

(a)

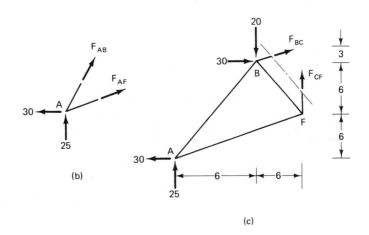

(b)

(c)

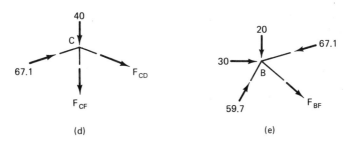

(d)

(e)

FIGURE 3.9

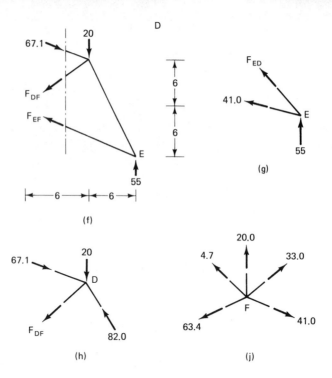

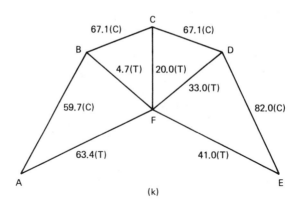

FIGURE 3.9 (Continued.)

Finally, F_{ED} is obtained by considering joint E in Fig. 3.9g.

$$\Sigma F_x = 0$$
$$0.894(41.0) + 0.447F_{ED} = 0$$
$$F_{ED} = -82.0 \text{ kN}$$

and F_{DF} is determined from joint D in Fig. 3.9h.

$$\Sigma F_y = 0$$
$$20 + 0.447(67.1) - 0.894(82.0) + 0.707F_{DF} = 0$$
$$F_{DF} = 33.0$$

The calculations are checked using an equation of vertical equilibrium at joint F shown in Fig. 3.9i.

$$\Sigma F_y = 0$$

$$4.7(0.707) + 33.0(0.707) + 20.0 - 63.4(0.447) - 41.0(0.447) = 0$$

$$0 = 0$$

A summary of the results is given in Fig. 3.9j.

3.7 STABILITY AND DETERMINACY OF TRUSSES

In Section 2.4 a determinate structure was defined as one for which the number of equations of equilibrium are equal to the number of unknown reactions. By comparison, it was stated that an indeterminate structure is one that possesses more unknown reactions than equations of equilibrium and an unstable structure is one that has an insufficient number of reactions to prevent motion. In extending these criteria to trusses we shall see that stability and determinacy of trusses depends not only on the number of reactions but also on the number and arrangement of individual members. A truss is externally stable and determinate if it possesses the proper number of reactions and internally stable and determinate if it has the correct number of bars arranged in a proper manner.

In applying the method of joints, we have seen that two independent equations of equilibrium can be written at each joint of a truss and that these two equations can be used to solve for two unknown forces at that joint. In other words, if a truss possesses j joints, we can write $2j$ independent equations and solve for $2j$ unknown forces. Although we did not solve for the reactions using joint equations in the preceding illustrative problems, it is certainly possible and sometimes even preferable to do so. If this procedure is employed, it follows that a truss is determinate if the number of member forces and reactions is equal to twice the number of joints. That is,

$$2j = m + r \tag{3.1}$$

where j = the number of joints, m = the number of members, and r = the number of reactions. Equation (3.1) also implies that a truss will be indeterminate if $2j < m + r$ and unstable if $2j > m + r$.

For a truss to be stable and determinate it is necessary that Eq. (3.1) be satisfied. However, by itself Eq. (3.1) is insufficient to ensure stablity and determinacy. In other words, Eq. (3.1) is a necessary but not sufficient condition for stability and determinacy. For a truss to be stable and determinate it must, in addition to satisfying Eq. (3.1), have both the correct number of reactions and the correct number of bars arranged in a proper manner.

To get a clear understanding of these concepts, let us consider the trusses in Fig. 3.10. According to Eq. (3.1), the trusses in Figs. 3.10a and 3.10d are stable and determinate while those in Figs. 3.10b and 3.10c are indeterminate and unstable, respectively.

The truss in Fig. 3.10a is externally stable and determinate because it has three

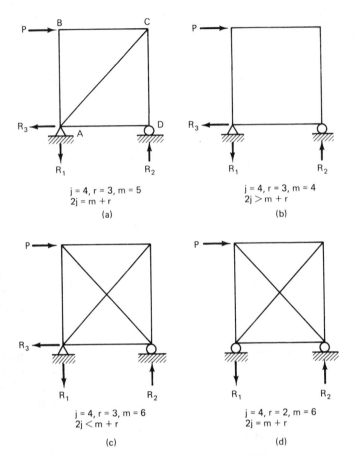

$j = 4, r = 3, m = 5$
$2j = m + r$
(a)

$j = 4, r = 3, m = 4$
$2j > m + r$
(b)

$j = 4, r = 3, m = 6$
$2j < m + r$
(c)

$j = 4, r = 2, m = 6$
$2j = m + r$
(d)

FIGURE 3.10

nonparallel, nonconcurrent reactions that suffice to prevent motion of the structure as a whole and that can be evaluated using the three equations of equilibrium available for a plane structure. The structure is also internally stable and determinate. As long as the bars are arranged to form one or more triangles, a rigid, stable system of members results. Furthermore, the bar forces can be evaluated using equations of equilibrium. For example, employing the method of joints, one could begin the analysis at either joint B or D, where there are only two unknowns, and then proceed to other joints to determine the remaining bar forces.

Like the truss in Fig. 3.10a, those in Figs. 3.10b and 3.10c are externally stable and determinate. The truss in Fig. 3.10b is, however, internally unstable. Because the bars are hinged to one another, the rectangle that they form can collapse. By comparison, the truss in Fig. 3.10c is internally indeterminate. That is to say the bar forces connot be evaluated using only equations of equilibrium. The presence of three unknown bar forces at each joint makes it impossible to use the method of joints. The method of sections is likewise inapplicable. Whereas the truss in Fig. 3.10b needs an interior diagonal to make it stable, the truss in Fig. 3.10c has one too many diagonals for determinacy.

Unlike the trusses in Figs. 3.10b and 3.10c, the one in Fig. 3.10d does satisfy Eq. (3.1). Nevertheless, it is neither stable nor determinate. It is externally unstable because the two vertical reactions are insufficient to prevent motion of the structure as a whole, and it is internally indeterminate because it has one more diagonal than necessary.

The foregoing examples illustrate the fact that Eq. (3.1) is a necessary but not sufficient condition for stability and determinacy. The trusses in Figs. 3.10b and 3.10c, which do not satisfy Eq. (3.1), are indeed unstable and indeterminate, respectively. However, the truss in Fig. 3.10d, although it satisfies Eq. (3.1), is neither stable nor determinate. The conclusion to be drawn from these examples is that a truss will be stable and determinate if in addition to satisfying Eq. (3.1) it is stable and determinate both internally and externally. This means that the bars must be arranged in the form of triangles or some other stable configuration and their number must be such that the bar forces can be determined using either the method of joints or the method of sections. In addition, there must be a sufficient number of reactions to prevent the structure from moving as a whole but not more than can be evaluated using the available equations of equilibrium.

For most trusses, stability and determinacy can be ascertained by using the foregoing principles. However, there are some trusses for which instability or indeterminacy becomes evident only after one attempts to analyze them.

3.8 COMPUTER SOLUTION

Whenever the solution of a problem involves the repeated application of a fixed procedure, a computer can be used advantageously to solve that problem. The analysis of a truss by the method of joints is such a problem. It requires that one sum components of bar forces, reactions and applied loads in the x- and y-directions at each joint in the truss.

In the latter part of the book we will consider the finite element method, which is a systematic procedure for analyzing not only trusses but all types of structures using the computer. However, at present let us develop a less general procedure, one whose sole purpose is to analyze a determinate truss with the aid of a computer. There are two ways this can be accomplished. On the one hand, we can write, by hand, an equation of vertical and horizontal equilibrium at every joint of the truss and then using a computer solve this set of simultaneous equations for the bar forces and reactions. Alternatively, we can let the computer not only solve the equations but also set them up. The latter procedure obviously necessitates the writing of a more involved program than does the former procedure. However, once the program has been written it obviates the need for a considerable amount of long hand calculation which is after all the primary purpose of using a computer.

To illustrate the first procedure, let us consider the truss in Fig. 3.11a. If we assume the reactions to be in the directions indicated and assume the bars to be stressed in tension, then the joints will have acting on them the forces shown in Fig. 3.11b. Applying equations of vertical and horizontal equilibrium to each of these joints leads to the following set of simultaneous equations.

$$
\begin{aligned}
0.6F_{AB} && + R_{AY} && &= 0 \\
0.8F_{AB} + F_{AC} && + R_{AX} && &= 0 \\
-0.6F_{AB} && - 0.6F_{BC} && &= 40 \\
-0.8F_{AB} && + 0.8F_{BC} && &= -20 \qquad (3.2) \\
&& + 0.6F_{BC} && + R_{CY} &= 0 \\
&& - F_{AC} - 0.8F_{BC} && &= 0
\end{aligned}
$$

Using the coefficients of the unknown bar forces and reactions and the values of the known applied loads as input, we may now employ any standard computer program

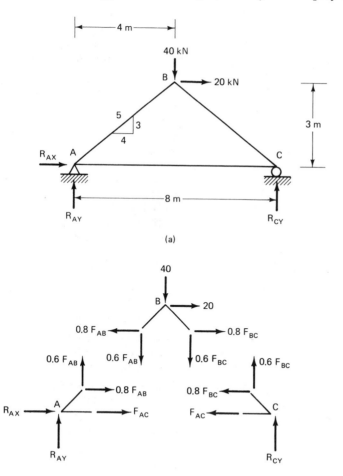

(a)

(b)   **FIGURE 3.11**

and solve this set of simultaneous equations for the bar forces and the reactions. If this is done, one obtains the following results.

$$F_{AB} = -20.83$$
$$F_{AC} = +36.67$$
$$F_{BC} = -45.83$$
$$R_{AY} = +12.50$$
$$R_{AX} = -20.0$$
$$R_{CY} = +27.50$$

Negative signs for the reactions indicate that they act in a direction opposite to the one assumed in the figure. For the bar forces positive signs denote tension and negative signs denote compression.

Let us now consider the second procedure, in which we let the computer set up the simultaneous equations as well as solve them. In other words, if we were analyzing the truss in Fig. 3.11, the computer would form the set of Eqs. (3.2) in addition to solving them. Before we consider the details of this procedure, let us note exactly what Eqs. (3.2) consist of.

1. Each row represents an equation of vertical or horizontal equilibrium at a joint. Since the truss has three joints, there are six equations.
2. The six unknowns on the left-hand sides of the equations include the three bar forces and the three reactions.
3. The two constants on the right-hand sides of the equations correspond to the two applied loads.

For the computer to be able to form the joint equilibrium equations, it must be given two sets of information. First the computer must be told which forces act in the x and y directions at each joint. In other words, one must tell the computer which nonzero terms appear in each equation. Second, the computer must be able to calculate the coefficients of the unknowns on the left-hand sides of the equations. There are two ways of dealing with the question of which forces act at any given joint. One can either proceed from joint to joint and simply note which members frame into the joint and which reactions and applied loads act on it, or one can consider the members, reactions, and applied loads one at a time and note which joints lie at their extremities. It turns out that the latter procedure is better suited to our task, and we will consequently use it. Thus we will tell the computer which forces act on each joint by considering each member, reaction, and applied load, one at a time, and note which joints lie at the extremities of each.

Having determined which forces appear in each joint equation, the computer must then calculate the coefficients of each unknown on the left-hand sides of the equations. In every instance the component of a force acting in either the x- or the y-direction at a joint is equal to the force multiplied by the cosine of the angle that the force makes with the x- or y-direction. In other words, the coefficients of the unknown forces in the equations are equal to the cosines of the angles between these forces and the x- and y-directions. These quantities, known as *direction cosines*, can be determined for the members by the computer if it is given the coordinates of the joints at the ends of the members.

We are now ready to consider the actual writing of the program that will set up and solve a set of joint equilibrium equations of the form

$$A_{11}F_1 + A_{12}F_2 \cdots + A_{1n}F_n = P_1$$
$$A_{21}F_1 + A_{22}F_2 \cdots + A_{2n}F_n = P_2$$
$$\vdots$$
$$A_{n1}F_1 + A_{n2}F_2 \cdots + A_{nn}F_n = P_n$$

where the F's represent the unknown bar forces and reactions, the A's represent the coefficients of these unknowns, and the P's are the known loads.

Using the truss in Fig. 3.12 as an illustrative example, we begin the procedure with the following steps.

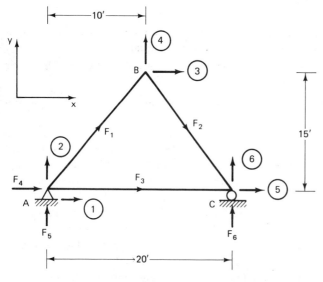

FIGURE 3.12

1. Decide on positive x- and y-directions. It simplifies matters if these are always chosen to the right and upward.

2. Number the members and the reactions. This establishes the identity of the unknown forces. For example, in our truss F_2 is the force in bar BC and F_5 is the vertical reaction at A.

3. Give each member a direction, making it in effect a vector. Thus member AB goes from A to B. This step is necessary to achieve a consistent sign convention for the direction cosines of the members.

4. Assume all bars to be in tension and all reactions in the positive x- and y-directions as indicated in the figure.

5. Number the joints using two numbers for each joint, one for the x-direction and one for the y-direction. Just as we have a single vector associated with each member, we now have two vectors associated with each joint. For example, the x-direction at joint B has been assigned the number 3 and the y-direction the number 4. The numbering of the joints is independent of the member numbering.

6. Choose a convenient origin and use it to determine the coordinates of the joints. Assign the x-coordinate of each joint to the x-vector and the y-coordinate to the y-vector. For example, if we assume the origin of the truss to be at A, the coordinates at B would be $c_3 = 10$ and $c_4 = 15$.

Let us now consider the flow-chart given in Fig. 3.13.

1. Read n, the number of simultaneous equations, and nm, the number of members.

2. Read P_1 through P_n, the values of the applied loads. The subscript on P refers

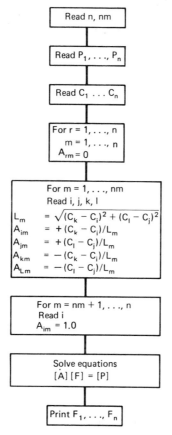

FIGURE 3.13

to the joint where P is applied. The sign of P depends on the direction of P and on whether the subroutine used to solve the simultaneous equations places the constants to the left or right of the equality sign.

3. Read c_1 through c_n, the coordinates of the joints.

4. Set all coefficients A_{rm} initially equal to zero.

5. Compute the nonzero coefficients A_{rm} of the bar forces by considering a single member at a time.

   (a) Let m, the number of the member, vary from 1 to nm, where $nm =$ the number of members in the truss.

   (b) Read for member m the joint numbers i, j, k, l at its extremities. As shown in Fig. 3.14, i and j correspond to the x- and y-vectors at the tail of the member and k and l represent the x- and y-vectors at the head of the member. For example, $i = 3$, $j = 4$, $k = 5$, and $l = 6$ for member 2 of the truss in Fig. 3.12.

   (c) Calculate L_m, the length of member m.

   (d) Calculate the four coefficients A_{rm} associated with member m by letting the subscript r take on successively the numbers read in for $i, j, k,$ and l in step (b). Fig. 3.14 demonstrates that a positive direction cosine leads to a positive bar-force component at the head of the member and to a negative bar-force component at the tail of the member. Conversely, if the direction

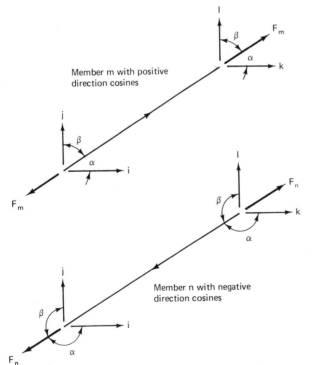

Member m with positive direction cosines

Member n with negative direction cosines

FIGURE 3.14

cosine of the member is negative, the signs of the bar-force components will be reversed.

6. Compute the nonzero coefficients A_{rm} of the reactions by considering one reaction at a time.
    (a) Let m, the number of the reaction, vary from $nm + 1$ to n.
    (b) Read for reaction m the joint number i at which the reaction is located.
    (c) Calculate the coefficient A_{im}. Since all reactions have been assumed in the positive x- and y-directions, A_{im} is always positive.

7. Using a subroutine, solve the simultaneous equations for the unknowns F_1 to F_n.

8. Print the values of F_1 to F_n.

PROBLEMS

3.1 to 3.7. Use the method of joints to determine all the bar forces.

3.1.

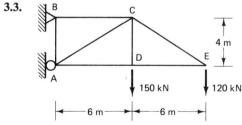

3.2.

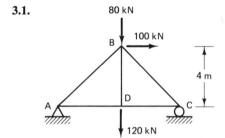

3.3.

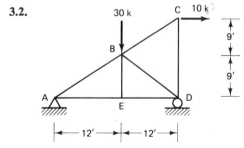

3.4.

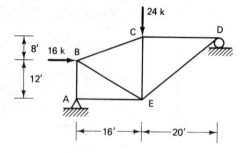

3.5.

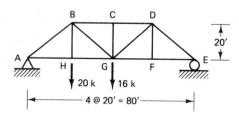

3.6.

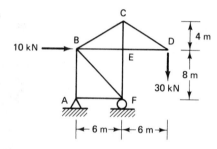

3.7.

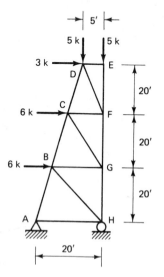

60

3.8 to 3.13. Use the method of sections to determine the bar forces for the indicated members.

3.8.   Bars *BC*, *BG*, and *HG*.

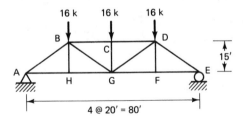

3.9.   Bars *BC*, *BG*, and *HG*.

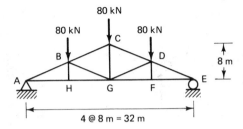

3.10.   Bars *BC*, *GC*, and *GF*.

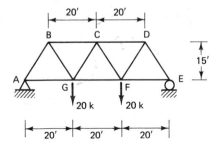

3.11.   Bars *BC*, *BD*, and *ED*.

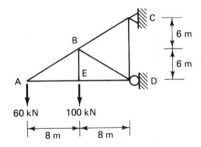

3.12.   Bars *BC*, *BG*, and *HG*.

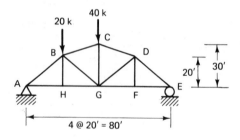

3.13.   Bars *BF*, *GF*, *AG*, and *CF*.

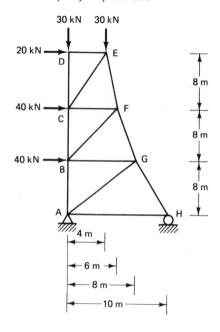

3.14 to 3.18. Use any method or combination of methods to determine all the bar forces.

3.14.

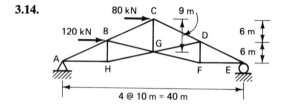

3.15.

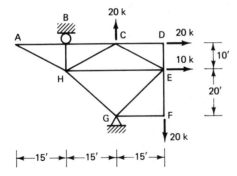

3.16.

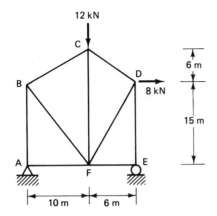

3.17.

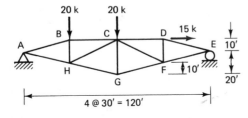

3.18.

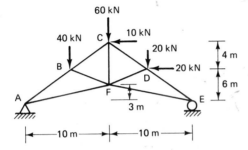

Offshore Drilling Platform.
(*Courtesy of Exxon Company, U.S.A.*)

4

Space Trusses

4.1 INTRODUCTION

Although the majority of structures are three-dimensional, it is usually possible to break them down into several plane components and analyze each of these by itself. For this reason, most of the material presented in this book deals with plane structures. There are, however, some three-dimensional structures, such as towers and reticulated domes, that must be analyzed in their entirety. This chapter presents an introduction to the analysis of such structures. Since we shall consider only the most elementary aspects of three-dimensional structures, we limit our attention to statically determinate three-dimensional trusses.

Three-dimensional structures are sometimes difficult to visualize, and geometrical properties such as lengths and directions are not always easy to determine. As a consequence, the analysis of a three-dimensional structure is usually considerably more involved than that of a two-dimensional structure, and the use of a computer program can be of great help. Nevertheless, some long-hand calculations are useful for introducing the basic principles involved in the analysis of three-dimensional structures.

4.2 CALCULATION OF REACTIONS

A three-dimensional truss is in equilibrium if the sum of all the forces acting on it is zero along any three mutually perpendicular axes and if the sum of the moments of these forces about each of these axes is also zero. These six conditions stated in equation form are

$$\sum F_x = 0, \qquad \sum F_y = 0, \qquad \sum F_z = 0$$
$$\sum M_x = 0, \qquad \sum M_y = 0, \qquad \sum M_z = 0$$

(4.1)

Since there are six independent equations of equilibrium, it is possible, in general, to evaluate six unknown reactions for a three-dimensional truss.

If we are to limit ourselves to trusses with only six reaction components, it is necessary that the supports be relatively uncomplicated. We will therefore assume that each support is one of the following three types:

1. A ball and socket joint that can resist translation in three perpendicular directions and can therefore apply forces to the structure in the x-, y-, and z-directions.

2. A support that is able to resist movement in only two directions and can therefore apply forces to the structure only in these two directions.

3. A support that can prevent motion only in one direction and can consequently apply a reaction only in that direction.

4.3 CALCULATION OF MEMBER FORCES

We will assume that the members of a space truss are connected to one another by ball and socket joints and that external loads are applied to the truss only at its joints. The bars of the truss will consequently be subjected to axial forces only.

We will determine the forces in individual members of the truss by using the method of joints. A joint in a three-dimensional truss can have acting on it a set of concurrent forces having components in the x-, y-, and z-directions. Such a joint will be in equilibrium if the sum of the forces in each of these three directions vanishes, that is, if

$$\sum F_x = 0, \qquad \sum F_y = 0, \qquad \sum F_z = 0 \tag{4.2}$$

One can thus write three equations of equilibrium and solve for three unknown bar forces at each joint in a three-dimensional truss.

Calculation of the bar forces requires that they be resolved into x-, y-, and z-components. As in the case of plane trusses, these components are obtained using the slopes of the bars. Figure 4.1 depicts a bar AB, of a three-dimensional truss, on which a force F_{AB} is acting. The x-component of the force is given by

$$F_{AB, x} = F_{AB} \cos \theta_x = F_{AB} \left(\frac{AC}{AB} \right) \tag{4.3}$$

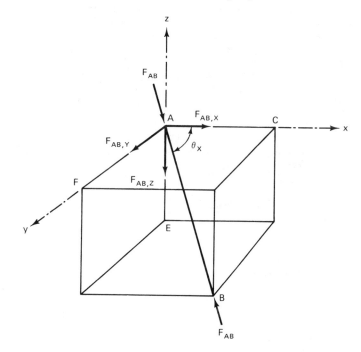

FIGURE 4.1

where θ_x is the angle between the member AB and its projection AC along the x-axis. Similarly,

$$F_{AB,Y} = F_{AB}\left(\frac{AF}{AB}\right), \qquad F_{AB,Z} = F_{AB}\left(\frac{AE}{AB}\right) \qquad (4.4)$$

The lengths of the projections of the member along the three coordinate axes are determined from the coordinates of the end points of the member. Having calculated these projections, one can then obtain the length of the member itself from

$$AB = \sqrt{(AC)^2 + (AF)^2 + (AE)^2} \qquad (4.5)$$

Example 4.1

As an illustration of the principles involved in analyzing a space truss, let us determine the reactions and the bar forces for the truss in Fig. 4.2.

As shown in Fig. 4.2a, the truss consists of six bars arranged in the form of a three-sided pyramid. The elevation and plan views of the truss are depicted in Figs. 4.2b and 4.2c. Two loads, acting in the z- and x-directions, respectively, are applied at point D, the apex of the structure. The base of the truss is restrained by three vertical reaction components, one each at points A, B, and C, two reaction components in the y-direction at points A and B, and a single reaction component in the x-direction at point C. The reactions are assumed to act in the directions shown in the figure.

We begin the analysis by calculating the reactions. There are six of them, and we have six independent equations to evaluate them. The vertical reaction at C, R_{CZ}, can be determined by taking moments about an x-axis passing through points A and B. Thus

$$\Sigma\, M_x = 0$$
$$(100)(4) - 8R_{CZ} = 0$$
$$R_{CZ} = 50$$

$$R_{CZ} = 50 \text{ kN} \uparrow$$

Similarly, a moment equation about a y-axis through point A allows one to obtain R_{BZ}.

$$\Sigma\, M_y = 0$$
$$100(4) - 200(12) - 50(4) - 8R_{BZ} = 0$$
$$R_{BZ} = -275$$

$$R_{BZ} = 275 \text{ kN} \downarrow$$

The value of R_{AZ} can now be determined by summing forces in the z-direction.

$$\Sigma\, F_z = 0$$
$$-100 + 50 - 275 + R_{AZ} = 0$$
$$R_{AZ} = 325$$

$$R_{AZ} = 325 \text{ kN} \uparrow$$

Summing forces in the x-direction allows us to calculate R_{CX}.

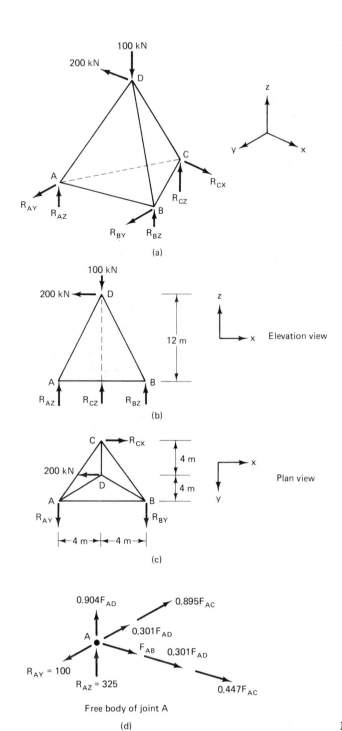

FIGURE 4.2

$$\Sigma F_x = 0$$

$$R_{CX} - 200 = 0$$

$$R_{CX} = 200$$

$$R_{CX} = 200 \text{ kN} \rightarrow$$

To determine R_{AY}, we sum moments about a z-axis through point B.

$$\Sigma M_z = 0$$

$$-200(4) + 200(8) - 8R_{AY} = 0$$

$$R_{AY} = 100$$

$$R_{AY} = 100 \text{ kN} \downarrow$$

Finally, to calculate R_{BY}, we sum forces in the y-direction.

$$\Sigma F_y = 0$$

$$R_{BY} + 100 = 0$$

$$R_{BY} = -100$$

$$R_{BY} = 100 \text{ kN} \uparrow$$

Having determined the reactions, we are now ready to calculate the bar forces. As indicated by Eqs. (4.3) and (4.4), these calculations require a knowledge of the lengths of the members, their x-, y-, and z-projections, and the ratios of these quantities. It is convenient to list these terms in tabular form as shown in Table 4.1.

TABLE 4.1

Member	Projection			Length	Ratio of projection to length			Force
	x	y	z		X/L	Y/L	Z/L	
AB	8	0	0	8.0	1.0	0	0	4.2
AC	4	8	0	8.94	0.447	0.895	0	232.6
AD	4	4	12	13.27	0.301	0.301	0.904	−359.5
BC	4	8	0	8.94	0.447	0.895	0	−214.1
BD	4	4	12	13.27	0.301	0.301	0.904	304.2
CD	0	4	12	12.65	0	0.316	0.949	−52.7

We begin the calculations by considering the free body of joint A shown in Fig. 4.2d. It should be noted that all bar forces have initially been assumed to be tension forces. Since the force in bar AD is the only force acting on the joint that has a z-component, summing forces in the z-direction immediately allows one to evaluate F_{AD}. Thus

$$\Sigma F_z = 0$$

$$0.904F_{AD} + 325 = 0$$

$$F_{AD} = -359.5 \text{ kN}$$

Next we write a force equation in the y-direction and determine the magnitude of F_{AC}.

$$\Sigma F_y = 0$$

$$100 - 0.301(-359.5) - 0.895F_{AC} = 0$$

$$F_{AC} = 232.6 \text{ kN}$$

Finally, summing forces in the x-direction gives F_{AB}.

$$\Sigma F_x = 0$$

$$F_{AB} + 0.301(-359.5) + 0.447(232.6) = 0$$

$$F_{AB} = 4.2 \text{ kN}$$

In a similar manner, using other joints, the remaining bar forces can be evaluated. The magnitudes of all bar forces are listed in the last column of Table 4.1.

Having familiarized ourselves with some of the basic principles involved in the analysis of a three-dimensional truss, let us now consider how we can simplify the analysis of such a structure with the aid of a computer. The most direct approach is to write three equations of equilibrium at every joint and then, using a computer, solve this set of simultaneous equations for both the reactions and the member forces.

Example 4.2

Determine the member forces and the reactions for the truss shown in Fig. 4.3.

The x-, y-, and z-projections of all bar lengths are listed in Table 4.2. Using this information, the lengths and the ratios of the projections to the lengths are calculated and noted in the table.

Assuming the reactions to be in the directions indicated in Fig. 4.3 and the bars to be in tension the simultaneous equations given in Table 4.3 are obtained. Each row in the table represents a force equation at one of the joints. For example, the first row contains the x-components of the forces acting on joint A. Similarly, the second and third row contain the y- and z-components of the forces acting on joint A. Each column in the table represents an unknown bar force, an unknown reaction, or a known load.

The results obtained by solving the equations in Table 4.3 are given in Table 4.4.

TABLE 4.2

Member	Projection			Length	Ratio of projection to length		
	x	y	z		X/L	Y/L	Z/L
AB	0	20	0	20	0	1.0	0
BD	50	0	0	50	1.0	0	0
DC	0	20	0	20	0	1.0	0
CA	50	0	0	50	1.0	0	0
AE	20	10	20	30	0.667	0.333	0.667
BE	20	10	20	30	0.667	0.333	0.667
DE	30	10	20	37.42	0.802	0.267	0.534
CE	30	10	20	37.42	0.802	0.267	0.534

TABLE 4.3

	AB	BD	DC	CA	AE	BE	DE	CE	R_{AZ}	R_{BZ}	R_{CZ}	R_{DZ}	R_{AX}	R_{BY}	R_{DX}	C
AX				1.0	0.667								1.0			
AY	−1.0				−0.333											
AZ					0.667				1.0							
BX		1.0				0.667										
BY	1.0					0.333								1.0		
BZ						0.667				1.0						
CX				−1.0				−0.802								
CY			−1.0					−0.267								
CZ								0.534			1.0					
DX		−1.0					−0.802								1.0	
DY			1.0				0.267									
DZ							0.534					1.0				
EX					−0.667	−0.667	0.802	0.802								
EY					0.333	−0.333	−0.267	0.267								20
EZ					−0.667	−0.667	−0.534	−0.534								40

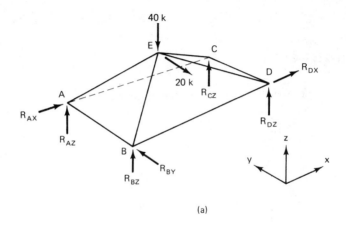

(a)

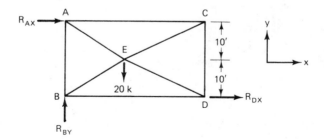

Elevation view

(b)

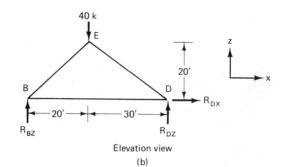

Plan view

(c)

FIGURE 4.3

TABLE 4.4

Reaction	Force (k)	Member	Force (k)
R_{AZ}	−8	AB	−4
R_{BZ}	32	BD	32
R_{CZ}	8	DC	4
R_{DZ}	8	CA	12
R_{BY}	20	AE	12
R_{AX}	−20	BE	−48
R_{DX}	20	DE	−15
		CE	−15

PROBLEMS

4.1 to 4.5. Determine the reactions and all the bar forces.

4.1.

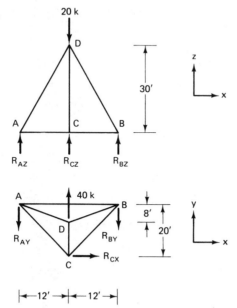

4.2.

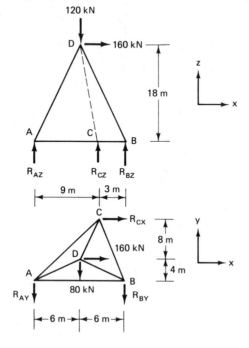

4.3.

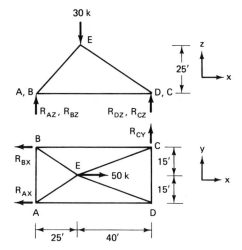

4.4.

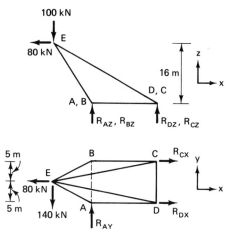

4.5.

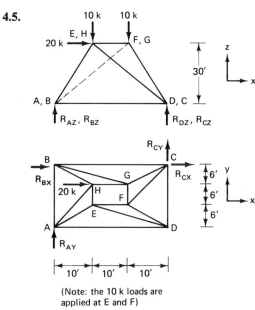

(Note: the 10 k loads are
applied at E and F)

5

Shear and Moment Diagrams for Beams and Frames

5.1 INTRODUCTION

Beams are relatively long and slender members that are usually loaded normal to their longitudinal axis. They transfer loads by developing a combination of bending and shear stresses. Some examples of beams are (1) the wooden joists that support the floors in single family dwellings, (2) the horizontal steel or concrete members that are used to support floors in large multistory buildings, and (3) the girders that support the roadway in small highway bridges. Beams are used to span distances ranging from several feet to approximately 200 feet. For very large spans they are uneconomical and other structural forms are employed.

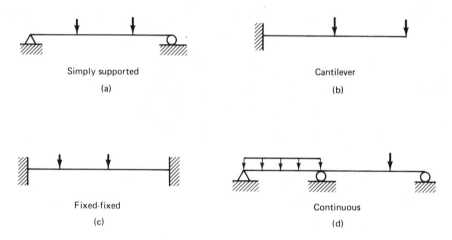

Simply supported

(a)

Cantilever

(b)

Fixed-fixed

(c)

Continuous

(d)

FIGURE 5.1

Beams are generally classified according to the manner in which they are supported. Some of the more common types are illustrated in Fig. 5.1. A *simply supported beam* is one that is supported by a hinge at one end and a roller at the other end (Fig. 5.1a). A *cantilever beam* is completely fixed at one end and free at the other end (Fig. 5.1b). If both ends of a beam are fixed it is referred to as a *fixed-fixed beam* (Fig. 5.1c). Sometimes a beam, in addition to being supported at its ends, is also supported at one or more intermediary points (Fig. 5.1d). Such a beam, if it is continuous over the intermediary supports, is called a *continuous beam*. Beams, like any other structure, can be determinate or indeterminate depending on whether or not there are enough equations of equilibrium to determine all the unknown reactions. For example, the beams in Figs. 5.1a and 5.1b are *determinate*. They each possess three unknown reactions, and there are three equations of equilibrium available for determining these reactions. By comparison, the beams in Figs. 5.1c and 5.1d are *indeterminate* because they each have more than three unknown reactions but there are still only three equations available for determining the reactions.

5.2 INTERNAL FORCES

The analysis of a beam involves calculating the stresses produced by the applied loads. The student may recall from the study of Mechanics of Materials that stresses are proportional to internal forces. To determine the stresses in a member, one must therefore know the magnitude of the internal forces. These forces are in turn obtained by passing sections through the member and considering the free body on either side of the section. For example, let us consider the beam in Fig. 5.2. If we are interested in the internal forces at a distance x from the left-hand support, we pass a section a–a through that point and consider the free bodies thus formed. Since either the free body to the left or to the right of the section can be used, let us consider the left-hand free body. The forces that the remainder of the beam exerts on this free body, to keep it in equilibrium, consist of a vertical force V and a couple M. The force V, which puts the free body in vertical equilibrium, is called the *shear*, and the couple M, which is necessary for moment equilibrium, is called the *bending moment*.

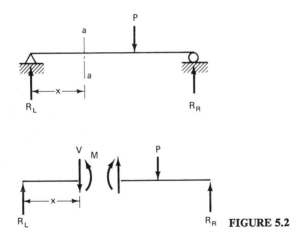

FIGURE 5.2

Unlike the axial force in the bar of a truss, which remains constant, the internal forces V and M in a beam vary along the span of the member. The shear and bending-moment diagrams, which are our main concern in this chapter, are devices for picturing how V and M vary. They are graphs of the magnitude of the shear and bending moment plotted along the span of the member. Once these diagrams have been plotted for a given beam, one can tell at a glance what the values of V and M are at any point along the beam and where the maximum values of V and M occur.

It is essential for the student to have a clear understanding of the difference between the internal couple that we refer to as the bending moment and the external moment about a point of all the forces acting on a member. The internal couple at any section is the moment required to keep the free body on either side of the section in equilibrium. It is also a measure of how much the beam is being bent. By comparison, the external moment about a point of all the forces acting both to the right and to the left of the point is an indication of whether or not the member will rotate. If the member is in equilibrium, this moment is equal to zero for all points.

5.3 SIGN CONVENTION

In order to be able to plot the variation of the shear and the bending moment along a member, it is necessary to adopt a sign convention for these quantities. The sign convention that we will use is illustrated in Fig. 5.3. As indicated, a shear force is considered to be positive if it produces a clockwise moment about a point in the free body on which it acts. Conversely, a negative shear force produces a counterclockwise moment about a point in the free body on which it acts. The bending moment is positive if it causes compression in the upper fibers of the beam and tension in the lower fibers. By comparison, a negative bending moment causes tension in the upper fibers and compression in the lower fibers. In other words, positive bending causes the upper surface of the beam to take on a concave shape, whereas negative bending causes a convex curve at the top of the beam.

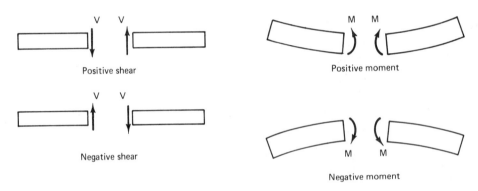

FIGURE 5.3

For a shear and moment sign convention to be useful, it must give the same results regardless of whether one considers the free body to the left or the right of the section. That the sign convention we have adopted satisfies this criterion is evident from Fig. 5.2. For this member both the shear and the bending moment are positive at section *a-a*, and this result can be obtained from either the left-hand or the right-hand free body.

5.4 SHEAR AND BENDING-MOMENT DIAGRAMS BY THE METHOD OF SECTIONS

The shear and bending-moment diagrams are graphical representations of the variation of the shear and bending moment along the span of the beam. It has already been demonstrated, in Fig. 5.2, that the shear and moment at any point along the beam can be obtained by passing a section through that point and by considering either of the two free bodies thus formed. If this process is repeated at regular intervals along the span of the member, curves of shear and moment versus distance can be plotted. For example, let us consider the beam in Fig. 5.4a. If sections are passed

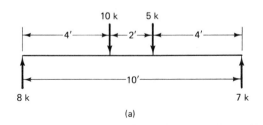

(a)

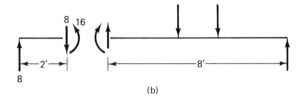

(b)

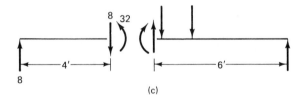

(c)

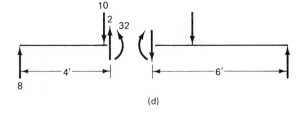

(d)

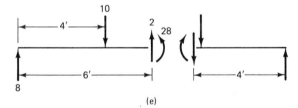

(e)   **FIGURE 5.4**

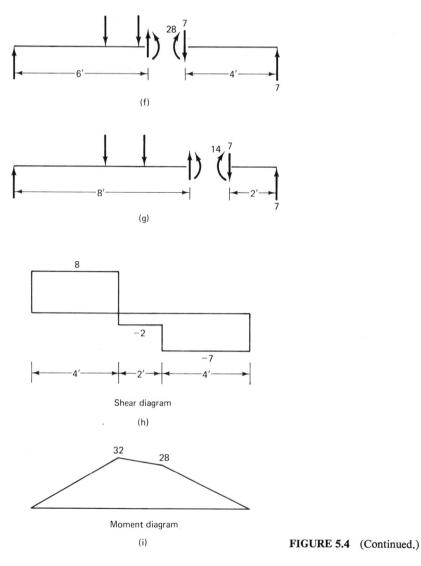

(f)

(g)

Shear diagram

(h)

Moment diagram

(i)

FIGURE 5.4 (Continued.)

through the beam at intervals of 2 ft, the free bodies shown in Figs. 5.4b through 5.4g are obtained. In each case the forces at the cut are obtained by putting the free body on which they act into vertical and moment equilibrium. Since the same internal forces act on both the free body to the right and to the left of a section, only one of the two need be considered. In each instance we have used the free body that involves less numerical work in the calculation of V and M. It should be noted that at 4 ft and 6 ft from the left end of the member, two separate sections are taken, one to the left of the concentrated load and one to the right of it. The reason for this is that the shear changes abruptly at a concentrated load and, while it is possible to calculate the shear just before or just after such a change, it is not possible to obtain its value at the point where it is changing.

Using the shears and moments obtained from the free bodies in Figs. 5.4b through 5.4g, we have plotted the diagrams in Figs. 5.4h and 5.4i. Although shear and moment diagrams can be obtained by cutting sections at regular intervals, as was done above, this procedure is as a rule too time consuming. Instead, a more efficient way of constructing shear and moment diagrams is to utilize other techniques in addition to the method of sections. In the following sections these techniques will be introduced and their use illustrated.

5.5 RELATIONSHIPS BETWEEN LOAD, SHEAR, AND BENDING MOMENT

There exist certain simple mathematical relationships between the load, the shear, and the bending moment in a beam, which can be extremely useful in the construction of shear and bending moment diagrams. For instance, let us consider the beam in Fig. 5.5a and concentrate our attention on the differential element of the beam shown in Fig. 5.5b. As indicated, the element has a shear V and a moment M acting on its left face and a shear $V + dV$ and a moment $M + dM$ acting on its right face. The quantities dV and dM represent the changes in the shear and moment that occur over the distance dx. All forces are shown acting in positive directions.

Writing an equation of equilibrium for forces in the vertical direction gives

$$V - (V + dV) - wdx = 0$$

from which

$$dV = -wdx \tag{5.1}$$

This expression states that the change in shear between two points along the member is equal to the load between the points. The minus sign indicates that a downward load, which we have assumed to be positive, results in an algebraic decrease in the shear.

A second relationship is obtained by summing moments about an axis through the left-hand face of the element. Thus

$$(V + dV)\,dx + \frac{w\,(dx)^2}{2} - (M + dM) + M = 0$$

If we neglect higher-order terms, this expression reduces to

$$\frac{dM}{dx} = V \tag{5.2}$$

or

$$dM = Vdx \tag{5.3}$$

The first of these two relations states that the slope of the moment diagram at any point is equal to the shear at that point, and the second relation states that the change

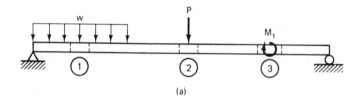

(a)

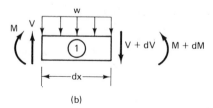

(b)

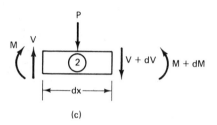

(c)

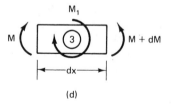

(d)

FIGURE 5.5

in moment between two points is equal to the area of the shear diagram between the points.

Let us now consider the element depicted in Fig. 5.5c and write an equation of vertical equilibrium for it. Thus

$$dV = -P \tag{5.4}$$

This expression states that the shear decreases by an amount equal to the concentrated load whenever one passes from the left to the right of such a load.

Similarly, by writing an equation of moment equilibrium for the element in Fig. 5.5d we obtain

84

$$dM = M_1 \tag{5.5}$$

which states that a couple results in an abrupt change of the bending moment, similar to the abrupt change in the shear caused by a concentrated load.

5.6 CONSTRUCTION OF SHEAR AND MOMENT DIAGRAMS

We now have available to us all the principles necessary for the efficient construction of shear and moment diagrams. There remains only the task of becoming proficient in the application of these principles. Thus let us consider the following illustrative examples.

Example 5.1

Construct the shear and bending-moment diagrams for the beam in Fig. 5.6a.

The first step is to calculate the reactions. The importance of carrying out this step correctly warrants using two equations of equilibrium to solve for the reactions and a third equation to check the results. Thus

$$\sum M_B = 0$$
$$20R_A - 10(14) - 20(8) = 0$$
$$R_A = 15 \text{ kips}$$

$$\sum M_A = 0$$
$$-20R_B + 20(12) + 10(6) = 0$$
$$R_B = 15 \text{ kips}$$

$$\sum F_y = 0$$
$$15 + 15 = 10 + 20$$

To construct the shear diagram for a beam with concentrated loads, one simply applies Eq. (5.4), which states that at each concentrated load the shear changes by an amount equal to the load. Equation (5.4) also implies that the shear remains constant between loads. If one starts at the left end of the beam and works toward the right, upward forces cause an increase in shear and downward forces a decrease in the shear. Using this procedure, the shear diagram in Fig. 5.6d is obtained.

The bending-moment diagram can be constructed either by passing sections through the member and by considering the free bodies thus formed or by making use of Eqs. (5.2) and (5.3). From the free body in Fig. 5.6b it is evident that the moment increases linearly from 0 to 90 kip-ft as x varies from 0 to 6 ft. The same information can also be obtained using Eqs. (5.2) and (5.3). In accordance with Eq. (5.2) the slope of the moment diagrams is equal to the shear. Since the shear is positive and constant between points A and C, the slope of the moment diagram must also be positive and constant between these points. In other words, the moment increases linearly with x. Equation (5.3) states that the change in moment between A and C is equal to the area of the shear diagram between A and C. The change in moment between these points is thus equal to 90 kip-ft. Further use of Eqs. (5.2) and (5.3) indicates that the shear and consequently the slope of the moment diagram is smaller between C and D than it was

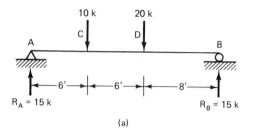

(a)

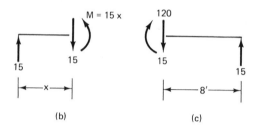

(b)   (c)

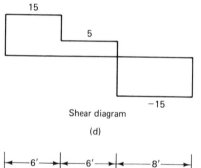

Shear diagram

(d)

Moment diagram

(e)   **FIGURE 5.6**

between A and C. It also indicates that the change in moment between C and D is equal to 30 kip-ft. The moment at D is thus equal to 120 kip-ft. This checks with the value of the moment at D obtained from the free body shown in Fig. 5.6c. Finally, the slope of the moment diagram between D and B is negative because the shear is negative between these points.

The following conclusions, regarding moment and shear diagrams, can be arrived at from the foregoing example. When a beam is subjected to concentrated loads,

86

1. The shear changes abruptly in value at each concentrated load and remains constant in value between loads.

2. The moment diagram consists of a series of straight lines whose slopes are positive or negative depending on whether the shear is positive or negative in the given region.

3. The maximum moment occurs where the shear changes sign.

Example 5.2

Construct the shear and moment diagrams for the beam in Fig. 5.7a.

The first step is to determine the reactions. Although the shear and moment diagrams must be drawn using the actual distributed load, in calculating the reactions one can replace the distributed load by a single concentrated force of 240 kN located 6 ft from the ends of the beam.

To draw the shear diagram, we make use of Eqs. (5.1) and (5.4), which state that the change in shear between two points is equal to the load between the points. The negative sign in the equations indicates that downward loads, which we consider to be positive, cause a decrease in the shear. Making use of Eq. (5.1), we find that the shear decreases by 80 kN between points A and B and by 160 kN between B and C. At B and C the reactions cause the shear to increase by 180 kN and 60 kN, respectively. Whereas concentrated loads, like the reactions at B and C, result in abrupt changes in the shear, distributed loads cause the shear to change gradually. In our case, the shear decreases by 20 kN per meter of member length because of the distributed load.

To construct the bending moment diagram, we use Eqs. (5.2) and (5.3) or free bodies like those in Figs. 5.7b and 5.7c. Equation (5.2) states that the slope of the moment diagram is equal to the shear. Accordingly the slope of the moment diagram, in Fig. 5.7e, decreases gradually from a zero slope at A to a negative slope at B. Then at point B there occurs an abrupt change in slope from negative to positive. Finally, between B and C the slope changes gradually from a positive slope at B to a negative slope at C.

To obtain values of the moment, we make use of Eq. (5.3), which states that the change in moment between two points is equal to the area of the shear diagram between the points. Thus the moment at B is equal to the area of the shear diagram between A and B, i.e., it is equal to -160 kN-m. Since the shear is negative between A and B, the moment decreases in magnitude as one goes from A to B. By comparison, the area of the shear diagram between B and D is positive, resulting in an increase of the moment between these points. To determine the value of the moment at D, one must first locate point D, the point where the shear is zero. This can be accomplished using the similar triangles that make up the shear diagram between B and C. Thus it can be determined that the distance between B and D is 5 m and that the moment at D is 90 kN-m.

Most of the information obtained above by using Eqs. (5.2) and (5.3) can be checked by considering free bodies such as those in Figs. 5.7b and 5.7c. For example, the free body in Fig. 5.7b indicates that the moment between A and B is a quadratic function and that its value at B is -160 kN-m. Similarly, the free body in Fig. 5.7c verifies that the moment at D is 90 kN-m.

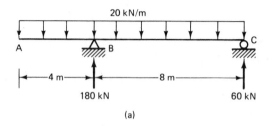

(a)

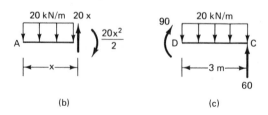

(b)   (c)

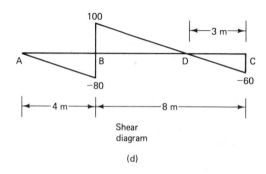

Shear
diagram

(d)

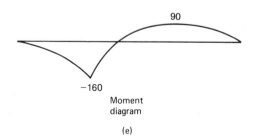

Moment
diagram

(e)

FIGURE 5.7

From the foregoing example we can conclude that for uniformly distributed loads the shear is a linear function of x and the bending moment is a quadratic function of x. In addition, the maximum values of the moment occur either where the shear is zero or where it changes sign.

Example 5.3

It is required to draw the shear and moment diagram for the beam in Fig. 5.8a.

First the reactions at A and B are determined. Then, using the principles outlined in the preceding pages, the diagrams in Figs. 5.8b and 5.8c are obtained.

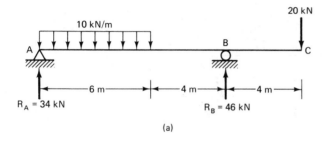

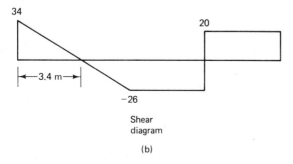

(a)

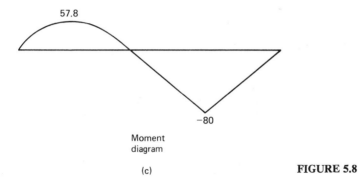

Shear
diagram

(b)

57.8

−80

Moment
diagram

(c)

FIGURE 5.8

Example 5.4

Construct the shear and moment diagram for the beam in Fig. 5.9a.

Equations (5.1) and (5.3) indicate that the shear in a beam is obtained by integrating the load, and the moment by integrating the shear. Since the given load varies linearly with x, the shear must be a quadratic function of x and the moment a cubic function of x.

To construct the shear diagram, we proceed along the same lines as in the previous examples. However, ordinates of the moment diagram are best obtained by using free bodies and not by attempting to calculate the areas of portions of the shear diagram. The latter procedure is helpful only when the shear diagram consists of triangular or rectangular areas.

The shear and moment at B are obtained using the free body of segment AB, shown in Fig. 5.9b. First, the intensity of the distributed load at B is calculated using similar triangles. Then equations of vertical and moment equilibrium of element AB are employed to determine the values of V and M at B.

89

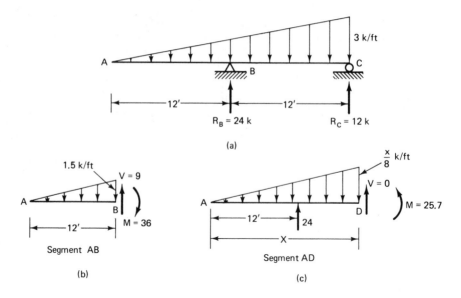

(a)

(b)   (c)

Segment AB

Segment AD

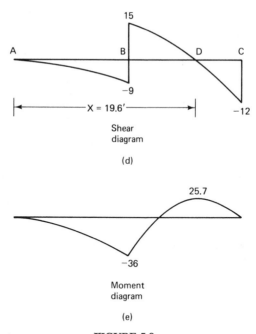

Shear
diagram

(d)

Moment
diagram

(e)

FIGURE 5.9

The maximum positive moment occurs at *D*, where the shear is zero. To obtain the magnitude of this moment, it is first necessary to locate point *D*. This can be accomplished by equating the upward and downward loads acting on segment *AD* in Fig. 5.9c. Thus

$$\frac{x}{8}\left(\frac{x}{2}\right) = 24$$

from which

$$x = 19.6$$

Knowing the value of x, we can determine both the magnitude and location of the resultant of the distributed load acting on segment AD, and hence the moment at D.

Example 5.5

Construct the shear and the moment diagram for member $ABCD$ of the structure in Fig. 5.10a.

To obtain the reactions at A and D, as well as the forces applied to member $ABCD$ at B, the structure is broken down into the three free bodies shown in Figs. 5.10b, 5.10c, and 5.10d. From the free body in Fig. 5.10d, the reaction at D and the shear at C are obtained. Then the free body in Fig. 5.10c is used to determine the forces that act at B on member ABC. Finally, the free body in Fig. 5.10b is used to calculate the reactions at A.

To determine the effect of the 90 kip-ft couple at B on the moment diagram of member $ABCD$, the free body in Fig. 5.10e is drawn. Since the moment to the left of B is -350 kip-ft and the 90 kip-ft couple is clockwise, it is evident that the moment to the right of B is equal to -260 kip-ft.

5.7 SHEAR AND MOMENT DIAGRAMS OF DETERMINATE, RIGID FRAMES

Flexural members do not occur only as isolated beams. Often several of them are connected to one another to form a frame. One of the most important of these is the *rigid frame*, which consists of flexural members connected to one another in such a manner that the members cannot rotate in relation to each other at the joints. As a consequence, moments as well as forces can be transferred from one member to another at their connection. By comparison, when two members are hinged to each other at a joint, the members can rotate in relation to one another and no moments can be transferred from one to the other.

Bending-moment and shear diagrams of rigid frames are easiest to construct if the frame is broken down into individual members and each of these is analyzed by itself. To see how this procedure is carried out, let us consider the following examples.

Example 5.6

Construct the shear and the moment diagrams for each member of the frame in Fig. 5.11a.

First we determine the reactions at A and D by applying the equations of equilibrium to the entire frame. Next we consider the three members out of which the frame is constructed and draw for each a free body a shear and a moment diagram. We begin with member AB, pictured in Fig. 5.11b. The free body of the member,

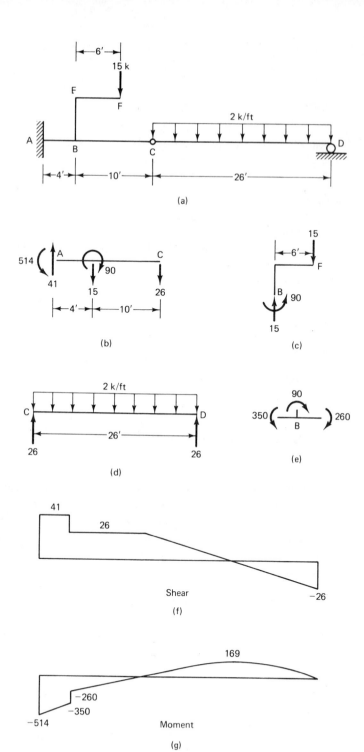

FIGURE 5.10

shown in the figure, extends from A to a section just below the joint at B. Thus, the 40 kN force acting on joint B does not appear on the free body we are considering. Knowing the magnitude of the forces acting at A and along the member, we can use equations of vertical, horizontal, and moment equilibrium to calculate the axial force, the shear, and the moment at the upper end of the member. This gives us a complete free body of member AB, from which shear and moment diagrams can readily be constructed.

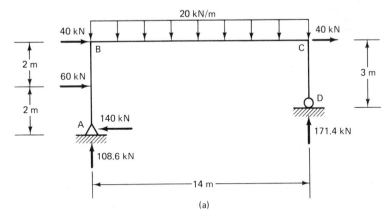

(a)

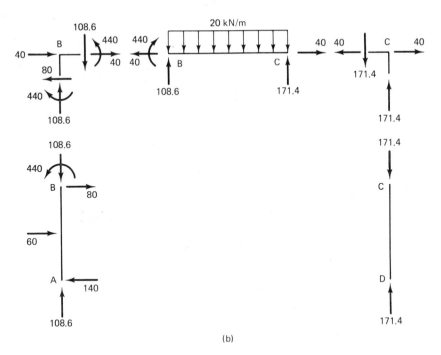

(b)

FIGURE 5.11

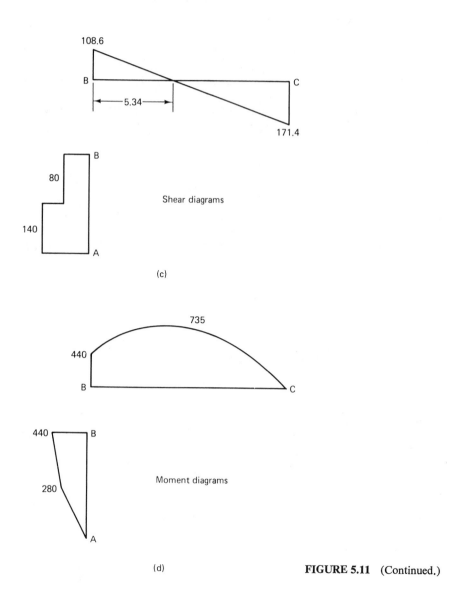

108.6

B ⌐‾‾‾‾‾‾‾‾‾‾‾‾‾‾‾‾‾C

|←—5.34—→|

171.4

B

80

140

A

(c)

Shear diagrams

735

440

B   C

440   B

280

A

(d)

Moment diagrams

FIGURE 5.11 (Continued.)

Before we can turn our attention to member *BC*, it is necessary to consider the free body of joint *B* shown in Fig. 5.11c. Starting with the forces at the lower end of the joint, which are obtained from member *AB* by applying the principle of action and reaction, and the external load of 40 kN applied directly to the joint, we obtain the forces at the right end of the joint using equations of equilibrium. It is necessary to consider a free body of a joint only when there are external loads applied directly to the joint. In the absence of such loads one can proceed directly from one member to the next.

We are now ready to deal with member *BC*, depicted in Fig. 5.11d. First we obtain the forces acting on the left end of the member from the free body of joint B. Then, using the equations of equilibrium, we determine the forces at the right end of the member needed to keep the member in equilibrium. From the free body thus obtained shear and moment diagrams are drawn.

The analysis of the frame is completed by drawing the free body of joint C, shown in Fig. 5.11e, and a free body of member CD, shown in Fig. 5.11f. Since member CD is subjected only to axial compression, the shear and moment are zero for the entire length of the member.

Since the members in a frame are not all horizontal, it is necessary at this time to replace the sign convention for bending that we have used in the past with a new one. Instead of referring to a moment as positive when it produces compression on the upper fibers of the member and negative when it produces compression on the bottom, we will from now on simply draw the moment diagram on the compression side of the member. This new procedure, in addition to being more versatile than the old one, produces the same moment diagram as the old one when the member is horizontal.

Example 5.7

Construct the shear and moment diagram for each member of the frame in Fig. 5.12a.

First the reactions at A and D are determined. Then a free body, a shear diagram, and a moment diagram are drawn for each of the three members. To draw the shear and moment diagram of member CD, it is necessary to resolve the forces acting on the ends of the member into components along and normal to the axis of the member. At C, this is accomplished by drawing a free body of the joint as shown in the figure.

Example 5.8

Construct shear and moment diagrams for each member of the frame in Fig. 5.13a.

As in previous examples, the reactions are determined first. Next free-body diagrams are drawn for each member, and finally, shear and moment diagrams are constructed.

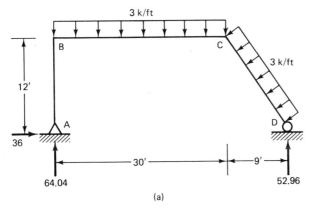

(a)   **FIGURE 5.12**

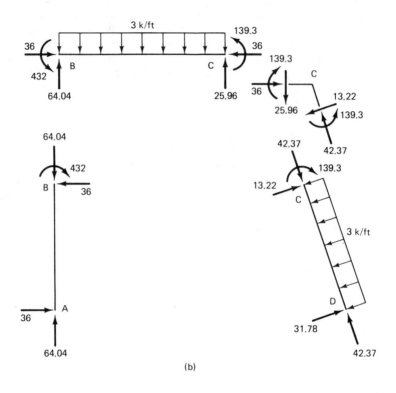

(b)

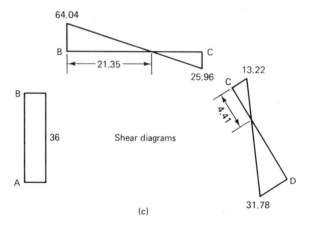

Shear diagrams

(c)

FIGURE 5.12 (Continued.)

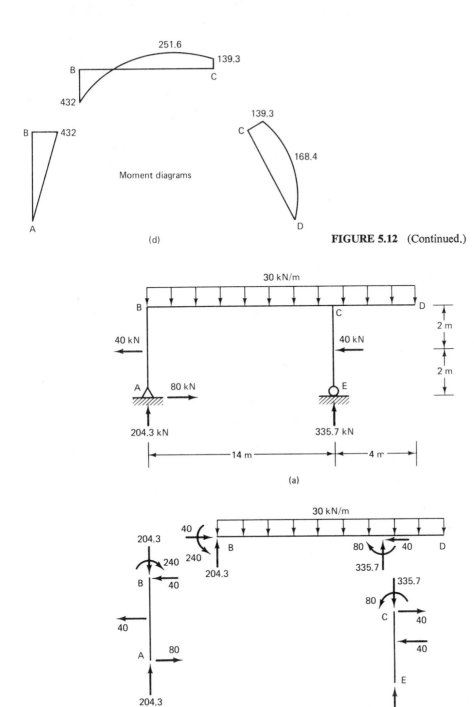

Moment diagrams

(d)

FIGURE 5.12 (Continued.)

(a)

(b)

FIGURE 5.13

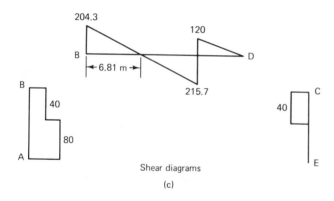

Shear diagrams

(c)

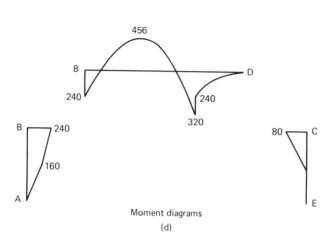

Moment diagrams

(d)

FIGURE 5.13 (Continued.)

PROBLEMS

5.1 to 5.27. Draw shear and moment diagrams.

5.1.

5.2.

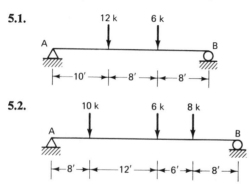

5.3.

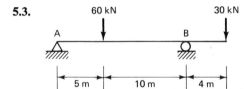

5.4.

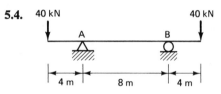

5.5.

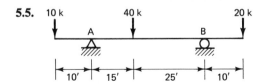

5.6.

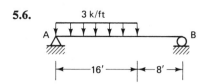

5.7.

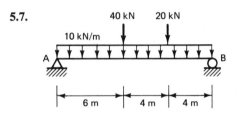

5.8.

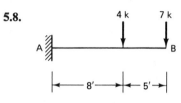

5.9.

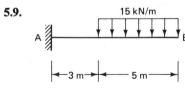

5.10.

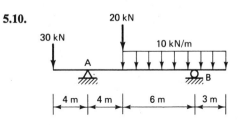

5.11.

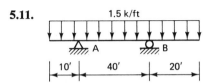

5.12.

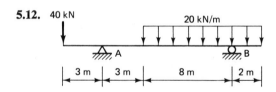

99

5.13.

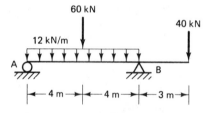

5.14.

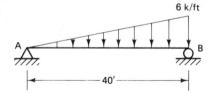

5.15.

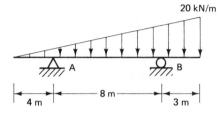

5.16.

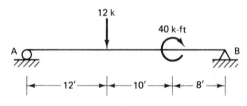

5.17.

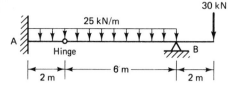

5.18.

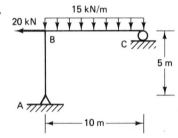

5.19.

5.20.

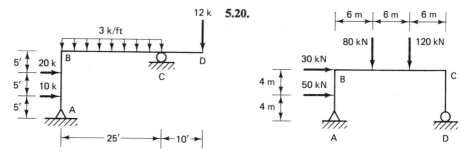

5.21.

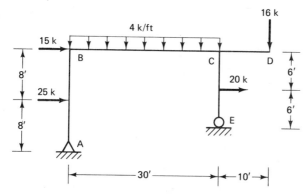

5.22.

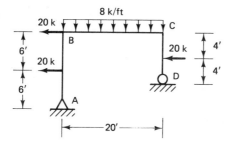

5.23.

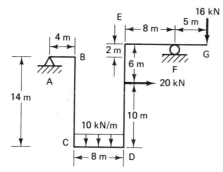

5.24.

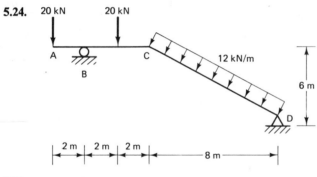

5.25.

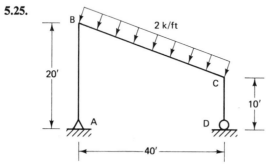

5.26.

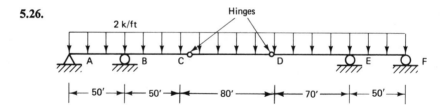

5.27.

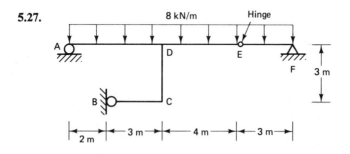

6

Deflections: Differential Equation

Newport Bridge, Newport, R.I. (*Courtesy of Bethlehem Steel Corporation.*)

6.1 INTRODUCTION

The structural engineer must be able not only to determine the internal forces and stresses that exist in a structure but also to calculate the magnitude of the deformations that occur. Just as it is necessary to ensure that allowable stresses are not exceeded, it is sometimes important to limit the size of the deflections that take place. Excessive deformations may prevent the structure from functioning properly. For example, if the members used to frame windows and doors in a building deform excessively, the free movement of the doors and windows will be hampered.

In some instances excessive deformations are undesirable not because they lead to malfunctioning of the structure but simply because they make people feel uncomfortable. A bridge may sag at its midpoint and still be structurally sound. However, such deflections tend to give an impression of weakness, which is undesirable. In the case of a bridge the problem is dealt with by cambering the structure; that is, the structure is given an initial upward curvature that masks the downward deflection. A similar problem arises in the upper stories of very tall buildings. Lateral deflections caused by strong winds are harmless from the point of view of structural safety. However, because they make the people occupying these spaces feel insecure, their magnitude must be limited.

A second and equally important reason for calculating deflections is that they play a vital role in the behavior of indeterminate structures. Since internal forces and deformations are coupled to one another in the equations that govern the behavior of these structures, both the deformations and the internal forces must be evaluated simultaneously. In other words, whereas one can obtain the internal forces for a determinate structure without calculating the deformations, the analysis of an indeterminate structure requires that both internal forces and deformations be calculated simultaneously.

In this and the following chapter we will study several different methods for calculating deflections. Since some of these are more suited to the analysis of one type of structure than another, it is important not only to learn the various methods but also to be able to choose the best method for a given task.

6.2 ELASTIC FORCE-DEFORMATION RELATIONSHIPS

The following is a summary of the elastic force-deformation relations for some common structural elements. The relations assume that (1) the material obeys Hooke's law (i.e., $\sigma \sim E\epsilon$), and (2) deformations are small compared to the original dimensions. In view of these assumptions, the behavior of the structural elements can be described by linear equations that lead to relatively simple numerical calculations. Since most engineering structures are designed so that the working stress does not exceed the proportional limit and since most engineering materials are very stiff below the proportional limit, the preceding assumptions are realistic for actual engineering structures. For example, steel having a working stress of 24 ksi and a modulus of 29×10^3 ksi will have at its working stress a strain equal to

104

$$\epsilon = \frac{\sigma}{E} = \frac{24}{29 \times 10^3} = 0.00082 \text{ in/in}$$

The elongation of a fiber of steel that is not stressed above its proportional limit will thus be less than $\frac{1}{1000}$ of its original length.

Axially Loaded Bars

The load-deflection relation of an axially loaded bar with constant cross-sectional area is given by

$$\delta = \frac{PL}{AE} \tag{6.1}$$

where δ is the elongation, L the length of the member, A its cross-sectional area, P the load, and E the modulus of elasticity.

Flexural Members

The moment-curvature relation for a flexural member subjected to pure bending is

$$\frac{1}{r} = \frac{M}{EI} \tag{6.2}$$

where r is the radius of curvature at any section and M, E, and I are the moment, the modulus of elasticity, and the moment of inertia at that section. Since

$$\frac{1}{r} = \frac{d\theta}{ds} = \frac{d^2y}{dx^2} \tag{6.3}$$

Eq. (6.2) can also be written in the form

$$\frac{d\theta}{ds} = \frac{M}{EI} \tag{6.4}$$

or

$$\frac{d^2y}{dx^2} = \frac{M}{EI} \tag{6.5}$$

The quantities appearing in Eqs. (6.2), (6.4), and (6.5) are defined in Fig. 6.1.

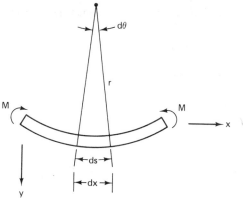

FIGURE 6.1

Torsional Members

The relation between the torque and the angle of twist for a torsional member is given by

$$\phi = \frac{TL}{GJ} \tag{6.6}$$

where ϕ is the angle of twist, J is the torsional constant, G the shear modulus, and T the torque. The form of J depends on the shape of the cross section.

Shear Panels

The load-deformation relation for a shear panel is

$$\delta = \frac{kVL}{AG} \tag{6.7}$$

where V is the shear force, A is the cross-sectional area of the edge on which V is acting, and L is the length of the side normal to the direction along which δ is measured, as shown in Fig. 6.2. The value of the factor k depends on the cross-sectional shape. It is equal to 1.2 for a rectangular section and 1.0 for a wide flange member.

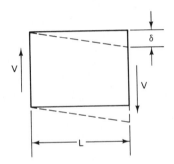

FIGURE 6.2

6.3 DIRECT INTEGRATION

It is possible, by integrating Eq. (6.5) twice, to obtain an expression for the deflection of a beam. To see how this is accomplished, let us consider the following illustrative examples.

Example 6.1

Obtain an expression for the deflection of the cantilever beam in Fig. 6.3. The stiffness EI of the beam is assumed to be constant.

If we take the origin of the coordinate system at the free end of the member, the moment at any section, a distance x from the origin, is

$$M = Px \tag{6.8}$$

Substitution of this expression into Eq. (6.5) gives

$$EI\frac{d^2y}{dx^2} = Px \tag{6.9}$$

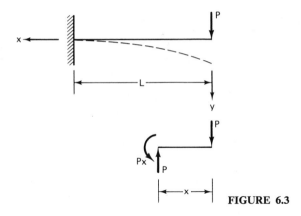

FIGURE 6.3

If we multiply both sides of the equation by dx and take the indefinite integral, we obtain

$$EI\frac{dy}{dx} = \frac{Px^2}{2} + C_1 \tag{6.10}$$

The constant of integration C_1 is present because we are not integrating between definite limits. A second integration of the equation leads to

$$EIy = \frac{Px^3}{6} + C_1 x + C_2 \tag{6.11}$$

To evaluate constants C_1 and C_2, we make use of the boundary conditions

$$y = \frac{dy}{dx} = 0 \quad @ \ x = L$$

Substitution of the boundary conditions in Eqs. (6.10) and (6.11) gives

$$C_1 = -\frac{PL^2}{2}$$

and

$$C_2 = \frac{PL^3}{3}$$

The expression for the deflection of the beam can now be written as

$$y = \frac{1}{EI}\left(\frac{Px^3}{6} - \frac{PL^2 x}{3} + \frac{PL^3}{2}\right) \tag{6.12}$$

To determine the deflection at the free end of the beam, let $x = 0$ in Eq. (6.12). This gives

$$y(0) = \frac{PL^3}{3EI}$$

Example 6.2

Obtain an expression for the deflection of the simply supported beam in Fig. 6.4. Choosing the left end of the beam as the origin of the coordinate system gives

$$M = \frac{Pbx}{L}, \quad 0 < x < a \tag{6.13}$$

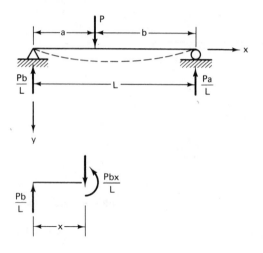

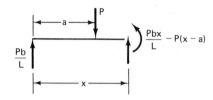

FIGURE 6.4

and

$$M = \frac{Pbx}{L} - P(x - a), \quad a < x < L \tag{6.14}$$

Since the moment is discontinuous at $x = a$, it is necessary to obtain separate solutions of Eq. (6.5) for the beam segments to the left and to the right of $x = a$.

$0 < x < a.$ Employing the expression in (6.13), we obtain

$$EI\frac{d^2y}{dx^2} = -\frac{Pbx}{L}$$

from which

$$EI\frac{dy}{dx} = -\frac{Pbx^2}{2L} + C_1 \tag{6.15}$$

and

$$EIy = -\frac{Pbx^3}{6L} + C_1x + C_2 \tag{6.16}$$

$a < x < L.$ Similarly, the expression in (6.14) gives

$$EI\frac{d^2y}{dx^2} = -\frac{Pbx}{L} + P(x - a)$$

$$EI\frac{dy}{dx} = -\frac{Pbx^2}{2L} + \frac{P(x - a)^2}{2} + C_3 \tag{6.17}$$

$$EIy = -\frac{Pbx^3}{6L} + \frac{P(x - a)^3}{6} + C_3x + C_4 \tag{6.18}$$

108

To evaluate the four constants C_1, C_2, C_3, and C_4, we make use of the boundary conditions

$$y = 0 \quad @ \ x = 0, \qquad y = 0 \quad @ \ x = L$$

and the conditions of continuity at $x = a$, namely, that the deflection y and the slope dy/dx given by equations (6.16) and (6.15) must be equal to the corresponding quantities given by Eqs. (6.18) and (6.17), at $x = a$.

Substitution of the first boundary condition in Eq. (6.16) gives

$$C_2 = 0$$

and equating the slopes given by Eqs. (6.15) and (6.17) at $x = a$ leads to

$$C_1 = C_3$$

Next we equate the deflection given by Eq. (6.16) to that given by Eq. (6.18) at $x = a$, which results in

$$C_4 = 0$$

Finally, applying the second boundary condition to Eq. (6.18) gives

$$C_3 = C_1 = \frac{Pb}{6L}(L^2 - b^2)$$

Substitution of the above constants into Eqs. (6.16) and (6.18) leads to the desired expressions for the deflection. Thus

$$y = \frac{Pbx}{6EIL}(L^2 - b^2 - x^2), \quad 0 < x < a$$

$$y = \frac{1}{EI}\left[\frac{Pbx}{6L}(L^2 - b^2 - x^2) + \frac{P(x-a)^3}{6}\right], \quad a < x < L$$

It is obvious from a comparison of Example 6.1 with Example 6.2 that the presence of discontinuities in the moment expression increases the complexity of the solution significantly. Thus, the solution for the deflection of a beam with three concentrated loads would require four separate equations and eight boundary conditions if we followed the procedure used in Example 6.2. Fortunately, this is not necessary. There exists a mathematical device called the *singularity function*, whose use removes the difficulties introduced by discontinuities in the moment expression. To see how singularity functions are used, let us consider the following example.

Example 6.3

Obtain an expression for the deflection of the beam in Fig. 6.4 using singularity functions.

As before, we choose the left end of the beam as the origin of the coordinate system. The moment at any section along the beam can be given by the single expression

$$M = \frac{Pbx}{L} - P\langle x - a \rangle \tag{6.19}$$

provided the function $\langle x - a \rangle$, called a *singularity function*, is defined as follows: $\langle x - a \rangle = 0$ if the expression inside the angle brackets is negative (i.e., if $x < a$) and $\langle x - a \rangle = x - a$ if the same expression is positive (i.e., if $x > a$).

With these facts in mind we substitute Eq. (6.19) into Eq. (6.5) and integrate twice. Thus

$$EI\frac{d^2y}{dx^2} = -\frac{Pbx}{L} + P\langle x - a\rangle$$

$$EI\frac{dy}{dx} = -\frac{Pbx^2}{2L} + P\frac{\langle x - a\rangle^2}{2} + C_1$$

$$EIy = -\frac{Pbx^3}{6L} + P\frac{\langle x - a\rangle^3}{6} + C_1x + C_2 \qquad (6.20)$$

The boundary conditions $y(0) = y(L) = 0$ give

$$C_2 = 0$$

and

$$C_1 = \frac{Pb}{6L}(L^2 - b^2)$$

Finally substituting these constants into Eq. (6.20) leads to

$$y = \frac{1}{EI}\left[\frac{Pbx}{6L}(L^2 - b^2 - x^2) + \frac{P\langle x - a\rangle^3}{6}\right]$$

Comparison of this result with the one in Example 6.2 indicates that the two are identical.

6.4 MOMENT-AREA THEOREMS

The moment-area method is a useful procedure for determining slopes and deflections in simple beams. Unlike the method of double integration, it does not give expressions for these quantities that are valid for the entire span of the beam. Instead, the moment-area method can only be used to determine the slope or the deflection at a specific point along the member. The moment-area method also differs from the method of double integration in that it makes use of the moment diagram instead of an analytical expression for the moment. As a consequence, discontinuities in the bending moment present no difficulties when applying the moment-area method.

The moment-area method involves the application of two theorems, which we will now introduce. Consider the beam, subjected to an arbitrary loading, shown in Fig. 6.5a. As a result of the loading, the beam deforms as indicated in Fig. 6.5b. If one takes the bending moment of this beam at any section and divides it by the stiffness EI at that section, one obtains the M/EI diagram shown in Fig. 6.5c. Let us focus our attention on an element ds of the beam. The tangents drawn to the ends of this element make an angle $d\theta$ with one another, as shown in Fig. 6.5b. According to Eq. (6.4), the relation between this angle and the corresponding ordinate of the M/EI diagram is

$$d\theta = \frac{M}{EI}\,ds$$

For small deformations, $dx \approx ds$ and

$$d\theta = \frac{M}{EI}\,dx \qquad (6.21)$$

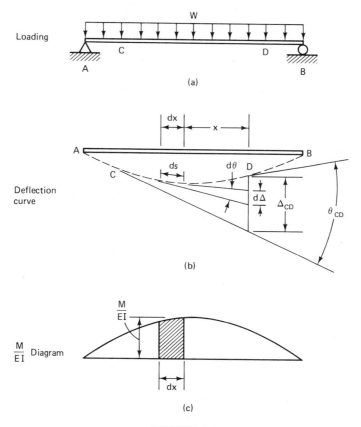

Loading

C

D

A

B

(a)

Deflection
curve

(b)

$\dfrac{M}{EI}$ Diagram

(c)

FIGURE 6.5

The change in slope between the ends of the element ds is thus equal to the area of the M/EI diagram corresponding to ds. To obtain the change in slope between two points such as C and D, which are a finite distance apart, one adds the incremental quantities $d\theta$ for all the elements between the points. Thus

$$\theta_{CD} = \int_C^D \frac{M}{EI} \, dx \qquad (6.22)$$

The first moment-area theorem given by Eq. (6.22) can be stated as follows:

Theorem 1. The change in slope between two points on the deflection curve of a beam is equal to the area of the M/EI diagram between these points.

It should be noted that this theorem gives only the change in slope between two points along a beam and not the value of the slope at a specific point.

To derive the second moment-area theorem, let us consider again the tangents drawn to the ends of element ds in Fig. 6.5b. The distance $d\Delta$ between these tangents,

111

measured along a vertical line through D, is given by

$$d\Delta = x\,d\theta \tag{6.23}$$

where x is the horizontal distance from D to the element ds. Substitution of Eq. (6.21) into Eq. (6.23) gives

$$d\Delta = x\,\frac{M}{EI}\,dx \tag{6.24}$$

The vertical distance $d\Delta$ is thus equal to the moment of the M/EI diagram corresponding to ds, about point D.

To obtain Δ_{CD}, the vertical distance from point D to a tangent drawn to C, we sum the incremental quantities $d\Delta$ for all the elements between C and D. Thus

$$\Delta_{CD} = \int_C^D x\frac{M}{EI}\,dx \tag{6.25}$$

This expression, referred to as the second moment-area theorem, can be stated as follows:

Theorem 2.   The vertical distance Δ, from a point D on the deflection curve of a beam to a tangent drawn to some other point C, is equal to the moment of the M/EI diagram between C and D about D.

The second moment-area theorem may at first be somewhat confusing. However, this confusion can be minimized if two facts are kept in mind:

1. The theorem does not give a beam deflection. Instead, it gives the vertical distance from one point on a beam to a tangent drawn to some other point.
2. The moment of the M/EI diagram is always taken about the point on the beam at which the vertical distance is measured.

6.5 APPLICATION OF THE MOMENT-AREA METHOD

The application of the moment-area method requires the determination of areas and centroids of bending-moment diagrams. To facilitate the evaluation of these quantities, the properties of several common areas are given in Fig. 6.6.

We intend to apply the moment-area method only to simple beams, for which the direction of the deflection and slope is best determined by sketching the deflected shape of the member. Therefore we will not develop a sign convention for the method.

Before considering some examples, it is worthwhile to note that the structures to which the moment-area method is usually applied can be subdivided into two groups: those for which the location of a horizontal tangent to the deflection curve is known and those for which we do not know where a horizontal tangent is located. Since the calculations are considerably easier to carry out when the location of a horizontal tangent is known, we will consider examples of this type first.

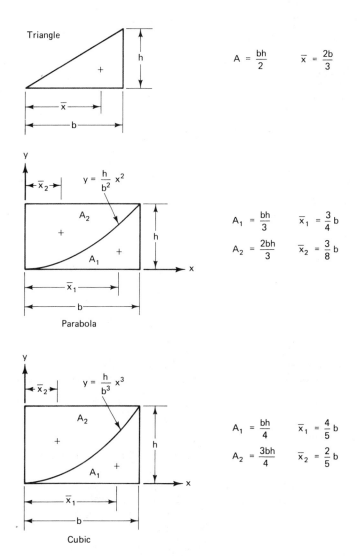

Triangle

$$A = \frac{bh}{2} \qquad \bar{x} = \frac{2b}{3}$$

$$y = \frac{h}{b^2} x^2$$

$$A_1 = \frac{bh}{3} \qquad \bar{x}_1 = \frac{3}{4} b$$

$$A_2 = \frac{2bh}{3} \qquad \bar{x}_2 = \frac{3}{8} b$$

Parabola

$$y = \frac{h}{b^3} x^3$$

$$A_1 = \frac{bh}{4} \qquad \bar{x}_1 = \frac{4}{5} b$$

$$A_2 = \frac{3bh}{4} \qquad \bar{x}_2 = \frac{2}{5} b$$

Cubic

FIGURE 6.6

Example 6.4

Determine the slope and deflection at the free end of the cantilever beam shown in Fig. 6.7. The beam is assumed to possess a constant stiffness *EI*.

The deflection curve of a cantilever beam always has a horizontal tangent at the fixed end of the member. The slope at any point along the member will therefore be equal to the change in slope between that point and the fixed end. Thus the slope at *B* is equal to the change in slope between *A* and *B*, which according to Theorem 1 is given by the area of the *M/EI* diagram between *A* and *B*. That is,

$$\theta_B = \left(\frac{PL}{EI}\right)\left(\frac{L}{2}\right) = \frac{PL^2}{2EI}$$

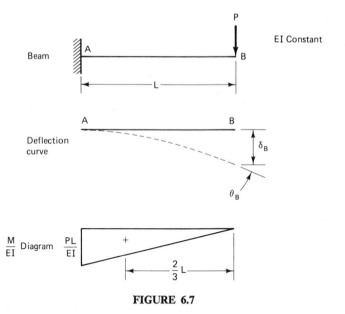

FIGURE 6.7

To determine the deflection at B, we use the second theorem, which states that the vertical distance from the deflection curve at B to the tangent drawn to A is equal to the moment of the M/EI diagram between A and B about B. Hence

$$\delta_B = \left(\frac{PL}{EI}\right)\left(\frac{L}{2}\right)\left(\frac{2L}{3}\right) = \frac{PL^3}{3EI}$$

Example 6.5

For the simply supported beam in Fig. 6.8, determine the deflection at midspan and the slope at A.

The deflection curve of a simply supported beam that is loaded symmetrically has a horizontal tangent at its midpoint. Hence the slope at A, for the beam in Fig. 6.8, is equal to the change in slope between A and B, which is equal to the area of the M/EI diagram between A and B. Thus

$$\theta_A = \frac{2}{3}\left(\frac{wL^2}{8EI}\right)\left(\frac{L}{2}\right) = \frac{wL^3}{24EI}$$

To calculate the deflection at B, we note that the distance d which can be determined using Theorem 2 is equal to δ_B. Applying Theorem 2, which states that d, the distance from C to a tangent drawn to B, is equal to the moment of the M/EI diagram between B and C about C, we obtain

$$\delta_B = d = \frac{2}{3}\left(\frac{wL^2}{8EI}\right)\left(\frac{L}{2}\right)\left(\frac{5L}{16}\right) = \frac{5wL^4}{384EI}$$

Example 6.6

It is required to determine the deflection at point B for the cantilever beam in Fig. 6.9. As indicated, the stiffness of the beam is equal to EI between B and C and $2EI$ between A and B.

114

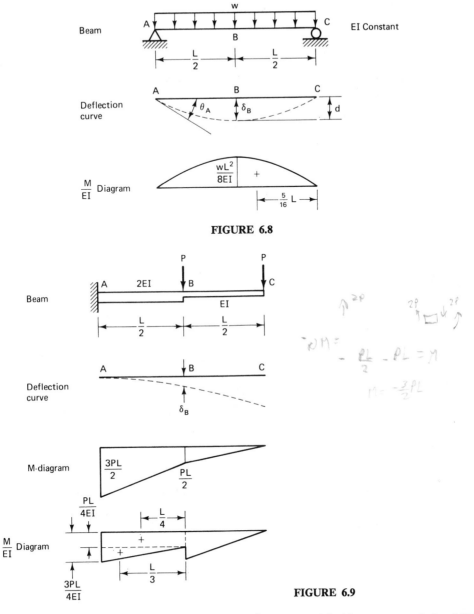

FIGURE 6.8

FIGURE 6.9

When dealing with a variable stiffness it is advisable to construct the M/EI diagram in two steps. First the ordinary moment diagram is drawn, and then the M/EI diagram is obtained by dividing the ordinates of the moment diagram by the appropriate stiffness.

According to Theorem 2, δ_B is equal to the moment of the M/EI diagram between A and B about B. As shown in the figure, the area of the M/EI diagram between

115

A and B is subdivided into a triangle and a rectangle. Thus

$$\delta_B = \frac{PL}{4EI}\left(\frac{L}{2}\right)\left(\frac{L}{4}\right) + \frac{PL}{2EI}\left(\frac{L}{4}\right)\left(\frac{L}{3}\right) = \frac{7PL^3}{96EI}$$

Example 6.7

Determine the slope at A and the deflection at B for the simply supported beam in Fig. 6.10.

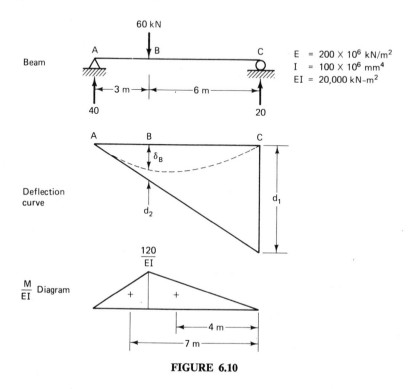

FIGURE 6.10

Since the loading is not symmetric, neither is the deflection curve. As a consequence, the location of the horizontal tangent is not known. Problems of this type are dealt with as follows. Drawing a tangent to the deflection curve at A and making use of similar triangles, one can write

$$\frac{\delta_B + d_2}{3} = \frac{d_1}{9}$$

from which

$$\delta_B = \frac{d_1}{3} - d_2$$

The quantities d_1 and d_2, which are the distances from C and B to the tangent drawn to A, are determined using the second theorem. Thus

$$d_1 = \left(\frac{180}{EI}\right)(7) + \left(\frac{360}{EI}\right)(4) = \frac{2700 \text{ kN-m}^3}{EI}$$

$$d_2 = \left(\frac{180}{EI}\right)(1) = \frac{180 \text{ kN-m}^3}{EI}$$

and

$$\delta_B = \frac{900 - 180}{EI} = \frac{720 \text{ kN-m}^3}{EI}$$

Substitution of the given values for E and I gives

$$\delta_B = \frac{720 \text{ kN-m}^3}{20,000 \text{ kN-m}^2} = 0.036 \text{ m} = 3.6 \text{ cm}$$

To obtain the slope at A, we note that slopes as well as deformations are assumed to be small. Hence

$$\theta_A = \tan \theta_A = \frac{d_1}{9}$$

$$\theta_A = \frac{2700 \text{ kN-m}^3}{9(20,000) \text{ kN-m}^3} = 0.015 \text{ radians}$$

Example 6.8

Determine the deflection at C for the beam in Fig. 6.11.

If a tangent is drawn to A, as shown in the figure, d_1 and d_3 can be obtained using Theorem 2. Also, from similar triangles, $d_2 = 1.5d_1$. Thus the desired deflection can be obtained from the relation

$$\delta_c = d_3 - d_2$$

$$d_1 = \frac{PL}{2EI}\left(\frac{L}{2}\right)\left(\frac{L}{3}\right) = \frac{PL^3}{12EI}$$

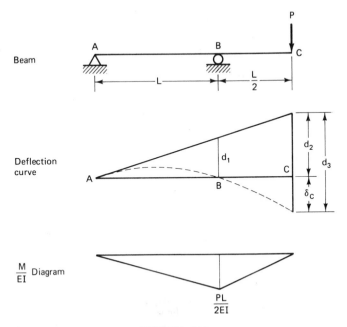

FIGURE 6.11

$$d_2 = 1.5d_1 = \frac{PL^3}{8EI}$$

$$d_3 = \frac{PL}{2EI}\left(\frac{L}{2}\right)\left(\frac{5L}{6}\right) + \frac{PL}{2EI}\left(\frac{L}{4}\right)\left(\frac{L}{3}\right) = \frac{PL^3}{4EI}$$

$$\delta_c = \frac{PL^3}{8EI}$$

Example 6.9

Determine the deflection at B for the beam in Fig. 6.12a.

If a beam is subjected to several loads, it is sometimes convenient to calculate deflections by making use of the principle of superposition. This procedure is applicable provided that the material obeys Hooke's law and deformations are small, i.e., provided that the behavior of the structure is linear.

In accordance with the principle of superposition the deflection at B due to both the distributed and concentrated load acting simultaneously (Fig. 6.12a) is equal to the sum of the deflections at B produced by the loads acting one at a time (Figs. 6.12b and 6.12c). Thus

$$\delta_{B1} = \frac{2}{3}\left(\frac{100}{EI}\right)(10)\left(\frac{5}{8}\right)(10) = \frac{4167 \text{ k-ft}^3}{EI}$$

$$d_2 = \frac{1}{2}\left(\frac{45}{EI}\right)(15)(10) + \frac{1}{2}\left(\frac{45}{EI}\right)(5)\left(\frac{10}{3}\right) = \frac{3750 \text{ k-ft}^3}{EI}$$

$$d_3 = \frac{1}{2}\left(\frac{30}{EI}\right)(10)\left(\frac{10}{3}\right) = \frac{500 \text{ k-ft}^3}{EI}$$

$$\delta_{B2} = 0.5d_2 - d_3 = \frac{1375 \text{ k-ft}^3}{EI}$$

$$\delta_B = \delta_{B1} + \delta_{B2} = \frac{5542 \text{ k-ft}^3}{EI}$$

Substitution of the given values for E and I and multiplication of the numerator by 1728 to obtain consistent units gives

$$\delta_B = \frac{5542(1728) \text{ k-in}^3}{29 \times 10^3(150) \text{ k-in}^2} = 2.20 \text{ in}$$

Example 6.10

Determine the deflection at C for the beam in Fig. 6.13a.

It is possible to apply the principle of superposition to the moment diagram of a beam as well as to its loads. In other words, just as one is able to break down the total loading into individual loads and consider them one at a time, it is possible to subdivide the moment diagram into several parts and consider these separately. For example, if we measure x from the left end of the member, the moment of the beam in Fig. 6.13a is given by

$$M = 45x, \quad 0 < x < 3$$

$$M = 45x - 40(x - 3), \quad 3 < x < 6$$

$$M = 45x - 40(x - 3) - \frac{10(x - 6)^2}{2}, \quad 6 < x < 12$$

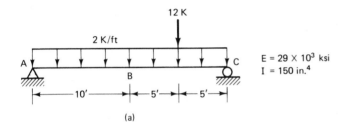

(a)

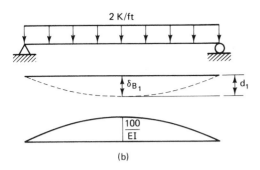

(b)

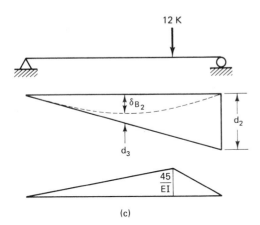

(c)

FIGURE 6.12

If the terms making up the above expressions are considered separately, then the moment diagrams shown in Fig. 6.13c result. Between A and B the moment consists of a single linear function, between B and C it is made up of the sum of two linear functions, and between C and D it consists of the sum of two linear functions and a quadratic function.

The deflection of C is now obtained by applying the moment-area theorem to the diagrams in Fig. 6.13c.

$$d_1 = \frac{1}{2}\left(\frac{540}{EI}\right)(12)(4) - \frac{1}{2}\left(\frac{360}{EI}\right)(9)(3) - \frac{1}{3}\left(\frac{180}{EI}\right)(6)\left(\frac{3}{2}\right) = \frac{7560}{EI}$$

119

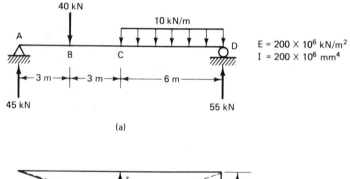

(a)

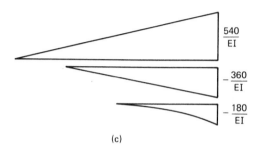

(b)

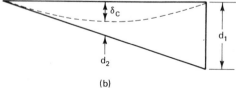

(c)

FIGURE 6.13

$$d_2 = \frac{1}{2}\left(\frac{270}{EI}\right)(6)(2) - \frac{1}{2}\left(\frac{120}{EI}\right)(3)(1) = \frac{1440}{EI}$$

$$\delta_c = 0.5d_1 - d_2 = \frac{2340}{EI}$$

or

$$\delta_c = \frac{2340 \text{ kN-m}^3}{40,000 \text{ kN-m}^2} = 0.0585 \text{ m} = 5.85 \text{ cm}$$

6.6 CONJUGATE-BEAM METHOD

In Section 6.3 it was demonstrated that the deflection of a beam can be obtained by integrating the expression for the bending moment twice. Unfortunately, this procedure becomes relatively involved whenever the bending moment cannot be written as a single continuous function for the entire beam. The conjugate-beam method, presented in this section, avoids this difficulty by substituting for the sometimes cumbersome process of integration the relatively simple procedure of constructing shear and moment diagrams. It is possible to do this because the drawing of the shear and moment diagrams is equivalent to the analytical process of integration.

120

Consider the differential beam element depicted in Fig. 6.14, on which are shown acting a distributed load w, shears V and $V + dV$, and moments M and $M + dM$. Summing forces in the vertical direction leads to

$$dV = -w\, dx$$

or

$$V = -\int w\, dx \tag{6.26}$$

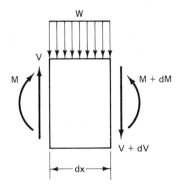

FIGURE 6.14

In addition, taking moments about the right end of the element and neglecting higher-order terms gives

$$dM = V\, dx$$

or

$$M = \int V\, dx \tag{6.27}$$

which in view of Eq. (6.26) can be rewritten as

$$M = \iint w\, dx\, dx \tag{6.28}$$

Equations (6.26) and (6.28) indicate that the shear in a beam can be obtained by integrating the load once and the moment by integrating the load twice. As a rule, structural engineers do not determine shears and moments by integrating analytic expressions for the loading. Instead, they draw shear and moment diagrams. In other words, the drawing of shear and moment diagrams can be substituted for the analytical process of integration. Let us now see how a similar substitution can be made in the calculation of slopes and deflections.

Since the curvature of a beam is proportional to the bending moment, the slope and deflection of the beam can be obtained by successively integrating the moment. Thus

$$\frac{d^2y}{dx^2} = \frac{M}{EI}$$

from which

$$\frac{dy}{dx} = \int \frac{M}{EI}\, dx \tag{6.29}$$

and

$$y = \iint \frac{M}{EI} \, dx \, dx \qquad (6.30)$$

It is our intention to replace the integrations indicated in Eqs. (6.29) and (6.30) by the drawing of shear and moment diagrams. To demonstrate how this is accomplished, let us consider the beam in Fig. 6.15. We begin by constructing the M/EI diagram of the beam. Next we introduce a new beam referred to as the *conjugate beam*. This beam has the same length as the real beam. However, instead of being loaded with the same load as the real beam, it is loaded with the M/EI diagram of the real beam. In other words, the conjugate beam is subjected to a distributed load whose intensity at any point is equal to the value of M/EI for the real beam at that point. We now proceed to construct the shear and moment diagrams of the conjugate beam.

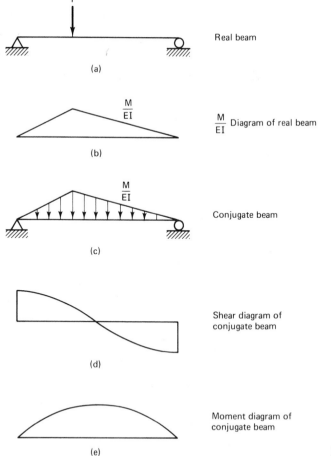

FIGURE 6.15

According to Eqs. (6.26) and (6.28) the drawing of V and M diagrams for a given loading is equivalent to two integrations of the loading. Thus the shear and moment diagrams of the conjugate beam represent one and two integrations, respectively, of the M/EI diagram of the real beam. But in view of Eqs. (6.29) and (6.30), integration of M/EI of the real beam gives the slope and deflection of the real beam. We thus conclude that the shear and moment diagrams of the conjugate beam represent the slope and deflection of the real beam.

The foregoing can be summarized as follows:

If we construct a conjugate beam whose span is equal to that of the real beam and whose load consists of the M/EI diagram of the real beam, then

1. The shear at any section of the conjugate beam is equal to the slope at that section in the real beam.
2. The moment at any section of the conjugate beam is equal to the deflection of the corresponding section of the real beam.

In the foregoing it has been tacitly assumed that the conjugate beam has the same support conditions as the real beam. Although this is true for a simply supported beam, it is not true for other boundary conditions. In general, the boundary conditions of the conjugate beam corresponding to a given real beam can be determined by making use of the relationships between the two beams stated in (1) and (2) above. If this is done, the results given in Table 6.1 are obtained. For example, if the real

TABLE 6.1

Real beam		Conjugate beam	
Support	Slope deflection	Shear moment	Support
Hinged	$y' \neq 0$ $y = 0$	$V \neq 0$ $M = 0$	Hinged
Fixed	$y' = 0$ $y = 0$	$V = 0$ $M = 0$	Free
Free	$y' \neq 0$ $y \neq 0$	$V \neq 0$ $M \neq 0$	Fixed
Continuous	$y' = $ continuous $y = 0$	$V = $ continuous $M = 0$	Hinge

beam has a fixed support, then $y = y' = 0$ at that support. At the corresponding support in the conjugate beam, $M = V = 0$, which defines a free end. Thus a fixed support in the real beam becomes a free end in the conjugate beam.

The conjugate-beam method, like the moment-area method, is primarily intended for the calculation of slopes and deflections in simple beams.

6.7 APPLICATION OF THE CONJUGATE-BEAM METHOD

Example 6.11

Determine the slope at A and deflection at B for the beam in Fig. 6.16.

The conjugate beam loaded with the M/EI diagram of the real beam is shown in Fig. 6.16b. Since the ends of the real beam are simply supported, so are the ends of the conjugate beam. The first step is to calculate the reactions of the conjugate beam.

$$R_A(L) - \frac{PL^2}{27EI}\left(\frac{7L}{9}\right) - \frac{2PL^2}{27EI}\left(\frac{4L}{9}\right) = 0$$

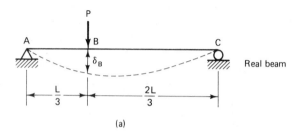

(a)

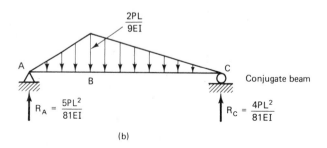

(b)

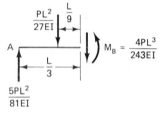

(c)

FIGURE 6.16

$$R_A = \frac{5PL^2}{81EI}$$

$$R_C = \frac{4PL^2}{81EI}$$

Since the slope in the real beam is equal to the shear in the conjugate beam, the slope θ_A at the left end of the real beam is equal to the left reaction of the conjugate beam. Thus

$$\theta_A = \frac{5PL^2}{81EI}$$

The deflection in the real beam is equal to the moment in the conjugate beam. Hence the deflection at B in the real beam is obtained by determining the moment at that point in the conjugate beam. Using the free body in Fig. 6.16c and summing moments about B gives

$$\frac{5PL^2}{81EI}\left(\frac{L}{3}\right) - \frac{PL^2}{27EI}\left(\frac{L}{9}\right) - M_B = 0$$

$$M_B = \frac{4PL^3}{243EI}$$

Thus

$$\delta_B = \frac{4PL^3}{243EI}$$

Note that both the slope and the deflection at specified points in the real beam can be determined simply by considering free bodies of the conjugate beam. It is unnecessary to draw either the shear or the moment diagram of the conjugate beam for this purpose. However, should the entire deflection curve of the real beam be required, then one would have to construct the bending-moment diagram of the conjugate beam.

Example 6.12

Determine the deflections at points B and C for the beam in Fig. 6.17a.

The conjugate beam loaded with the M/EI diagram of the real beam is shown in Fig. 6.17b. In accordance with Table 6.1, the conjugate beam is fixed at C and free at A. To calculate the reactions of the conjugate beam, the loading is divided into three parts as shown in the figure. From vertical equilibrium,

$$R_C = \frac{7wL^3}{48EI}$$

and from moment equilibrium,

$$M_C = \frac{wL^3}{16EI}\left(\frac{3L}{4}\right) + \frac{wL^3}{16EI}\left(\frac{5L}{6}\right) + \frac{wL^3}{48EI}\left(\frac{3L}{8}\right)$$

$$M_C = \frac{41wL^4}{384EI}$$

Since the deflection in the real beam is equal to the moment in the conjugate beam,

$$\delta_C = \frac{41wL^4}{384EI}$$

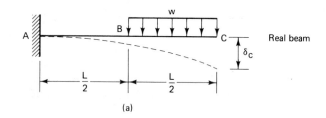

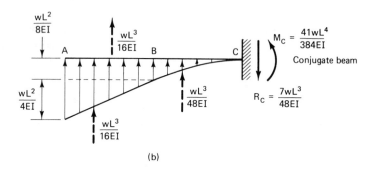

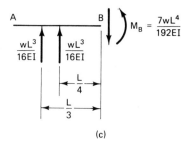

FIGURE 6.17

To obtain the deflection at B, we use the free body shown in Fig. 6.17c. Thus

$$M_B = \frac{wL^3}{16EI}\left(\frac{L}{3}\right) + \frac{wL^3}{16EI}\left(\frac{L}{4}\right) = \frac{7wL^4}{192EI}$$

and

$$\delta_B = \frac{7wL^4}{192EI}$$

Example 6.13

Calculate the deflection at C for the beam in Fig. 6.18a.

In accordance with Table 6.1, the conjugate beam shown in Fig. 6.18b has a simple support at A and a hinge at B and is fixed at C. To obtain the left reaction of the conjugate beam, we consider the free body to the left of B, shown in Fig. 6.18c.

126

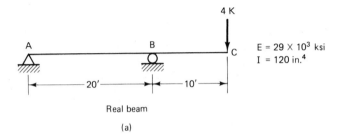

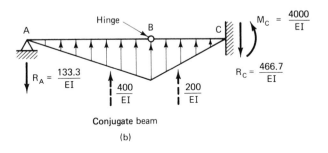

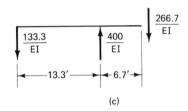

FIGURE 6.18

$$R_A(20) - \frac{400}{EI}(6.77) = 0$$

$$R_A = \frac{133.3}{EI}$$

The reactions at C can now be determined from a free body of the entire conjugate beam. Thus

$$R_C = \frac{466.7}{EI}$$

$$M_C = \frac{4000}{EI}$$

and

$$\delta_C = \frac{4000 \text{ kip-ft}^3}{EI}$$

Substituting for EI and multiplying the numerator by 1728 to obtain consistent units gives

$$\delta_C = \frac{4000(1728)}{29 \times 10^3(120)} = 1.99 \text{ in}$$

PROBLEMS

6.1 to 6.6. Use the method of double integration to obtain an expression for the deflection.

6.1.

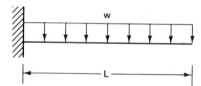

6.2.

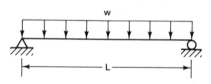

6.3.

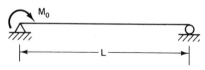

6.4.

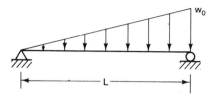

6.5.

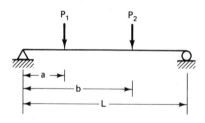

6.6.

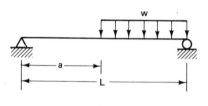

6.7 to 6.24. Use the moment-area method to calculate the indicated quantities.

6.7. Find δ_B, δ_C, θ_B, θ_C.

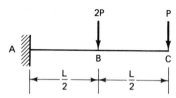

6.8. Find δ_B, θ_B.

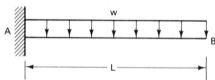

6.9. Find $\delta_B, \delta_C, \theta_C$; $E = 200 \times 10^6$ kN/m², $I = 20 \times 10^6$ mm⁴.

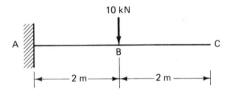

6.10. Find δ_C, θ_C; $E = 30 \times 10^3$ ksi, $I = 70$ in⁴.

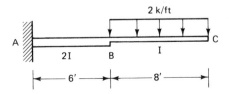

6.11. Find δ_B, θ_A.

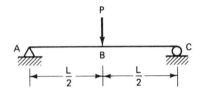

6.12. Find δ_B, θ_A; $E = 200 \times 10^6$ kN/m², $I = 50 \times 10^6$ mm⁴.

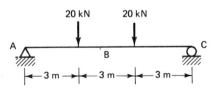

6.13. Find δ_B, θ_A.

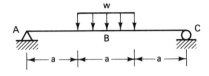

6.14. Find δ_A, δ_C; $E = 30 \times 10^3$ ksi, $I = 120$ in⁴.

6.15. Find δ_A, δ_C.

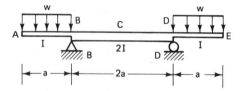

6.16. Find δ_B; $E = 200 \times 10^6$ kN/m², $I = 20 \times 10^6$ mm⁴.

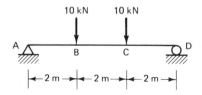

6.17. Find δ_B.

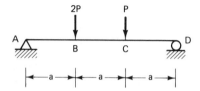

6.18. Find δ_D; $E = 30 \times 10^3$ ksi, $I = 144$ in⁴.

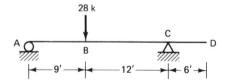

6.19. Find δ_C; $E = 200 \times 10^6$ kN/m², $I = 200 \times 10^6$ mm⁴.

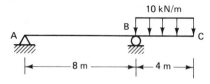

6.20. Find δ_C.

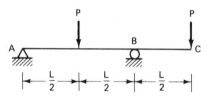

6.21. Find δ_A, δ_C; $E = 30 \times 10^3$ ksi, $I = 10$ in⁴.

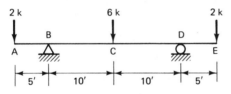

6.22. Find δ_B; $E = 30 \times 10^3$ ksi, $I = 24$ in⁴.

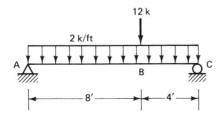

6.23. Find δ_C.

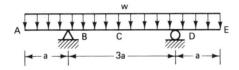

6.24. Find δ_B; $E = 200 \times 10^6$ kN/m², $I = 150 \times 10^6$ mm⁴.

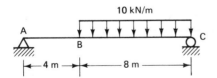

6.25 to 6.34. Use the conjugate-beam method to calculate the indicated quantities.

6.25. Find δ_B, θ_A.

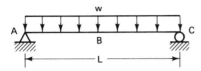

6.26. Find δ_B, θ_A; $E = 200 \times 10^6$ kN/m², $I = 50 \times 10^6$ mm⁴.

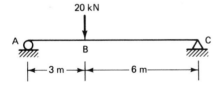

6.27. Find δ_B, θ_A.

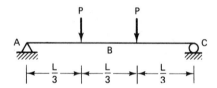

6.28. Find δ_C; $E = 30 \times 10^3$ ksi, $I = 24$ in^4.

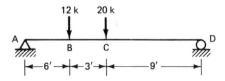

6.29. Find δ_B, θ_B.

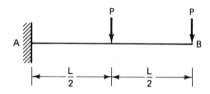

6.30. Find δ_C; $E = 200 \times 10^6$ kN/m^2, $I = 40 \times 10^6$ mm^4.

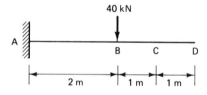

6.31. Find δ_B, θ_B; $E = 30 \times 10^3$ ksi, $I = 120$ in^4.

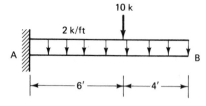

6.32. Find δ_C.

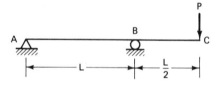

6.33. Find δ_C; $E = 30 \times 10^3$ ksi, $I = 80$ in^4.

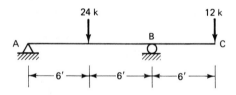

6.34. Find δ_A, δ_C; $E = 200 \times 10^6$ kN/m^2, $I = 50 \times 10^6$ mm^4.

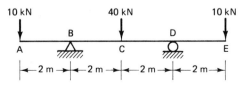

7

Deflections: Energy Methods

7.1 INTRODUCTION

In the previous chapter we calculated deflections using methods whose basis is the differential equation of the member. It will now be shown that deflections can also be obtained by using the principle of conservation of energy. This principle can be applied by using either the concept of real work or the concept of virtual work. Of the two approaches, the latter leads to a method that is far more versatile and consequently more useful than the former. However, in order fully to understand the ideas involved, we will consider both procedures.

7.2 PRINCIPLE OF CONSERVATION OF ENERGY

If a set of external loads are applied to a deformable structure, the points of application of the loads move and each of the members or elements making up the structure becomes deformed. In this situation both the external loads acting on the structure as a whole and the internal forces acting on the individual elements of the structure perform work. Applying the law of conservation of energy to such a system, one can state that the work performed on the structure by the external loads is equal to the work performed on the elements of the structure by the internal forces. This principle can be stated as

External Work = Internal Work

Since the work performed on the elements of a structure by the internal forces causes the elements to become deformed, the internal work is sometimes referred to as *strain energy*. The principle of conservation of energy is then stated in the form

External Work = Strain Energy

In a structural system the external work will be equal to the internal work only if the loads are applied gradually and no acceleration occurs and if the strains in the system remain elastic. If the system does accelerate, some of the external work will be transformed into kinetic energy; and if inelastic behavior occurs, energy will be lost in the form of heat. The above principle is thus applicable only to elastic systems subjected to static loads.

Before we apply the principle of conservation of energy, let us consider the concept of work. For the purposes of structural mechanics, *work* is defined as the product of a force and the displacement, in the direction of the force, through which the force acts. If the force remains constant in magnitude throughout the entire displacement, the work is simply equal to the product of the force and the total displacement.

$$W = P\Delta \tag{7.1}$$

However, in many instances the force does not remain constant. In this case the total work is obtained by summing small increments of work for which the force can be assumed to remain constant. Thus

$$W = \int P \, d\Delta \tag{7.2}$$

Let us now determine the work performed by force P_F in stretching a bar an amount Δ_F, as shown in Fig. 7.1a. If the load is applied in a static manner (that is, both the load and the resulting deformation increase together from zero to their final value) and if the material obeys Hooke's law, then the load-deflection curve will be as depicted in Fig. 7.1b. Since the applied load does not remain constant, we use Eq. (7.2) to calculate the total work. Thus

$$W = \int dW = \int_0^{\Delta_f} P \, d\Delta$$

Substituting $k\Delta$ for P

$$W = k \int_0^{\Delta_f} \Delta \, d\Delta = \frac{k\Delta_f^2}{2}$$

or, since $P_f = k\Delta_f$,

$$W = \frac{P_f \Delta_f}{2} \tag{7.3}$$

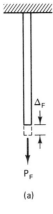

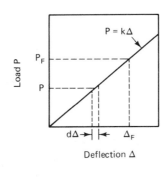

(a)   (b)   **FIGURE 7.1**

Thus the work performed by a static load applied to a structural system whose material obeys Hooke's law is equal to one half the product of the load and the corresponding deflection. Geometrically, this is the area under the load-deflection curve.

Similarly, it can be shown that the work due to a moment is equal to one half the product of the moment and the corresponding rotation. That is,

$$W = \frac{M\theta}{2} \tag{7.4}$$

and the work resulting from a torque is

$$W = \frac{T\phi}{2} \tag{7.5}$$

where ϕ is the angle of twist.

7.3 METHOD OF REAL WORK

To see how the principle of conservation of energy can be used to determine the deflection of a structure, let us consider the beam in Fig. 7.2a and calculate the deflection δ under the load P. It is assumed that P is applied gradually and that the material out of which the beam is made remains elastic.

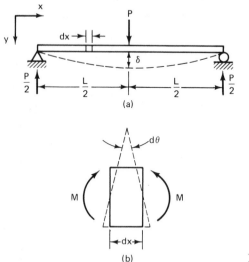

(a)

(b)   **FIGURE 7.2**

In order to apply the principle of conservation of energy, we must determine both the external and the internal work. The external work consists of the total work performed by all the external forces. Since the supports do not move in the vertical direction as the structure deforms, the reactions at A and C perform no work and only P contributes to the external work. Thus, in accordance with Eq. (7.3),

$$W_{\text{EXT}} = \frac{P\delta}{2} \tag{7.6}$$

Although the internal work theoretically consists of the total work due to all the internal forces, it is sometimes possible in certain structures to neglect the effects of some of the internal forces compared to that of others. In a truss only axial forces exist, and there is consequently no problem in deciding that they must be used to calculate the internal work. However, in a beam the internal forces consist of both shears and moments. Since it has been shown that the work performed by the shear forces is negligible compared to that due to the moments, provided the span of the beam is at least three times its depth, it is customary to use only the moments to calculate the internal work for ordinary beams.

Figure 7.2b depicts an element dx of the beam we are considering. As the beam bends, the cross sections of the element, on which the internal moments act, rotate relative to one another by an angle $d\theta$. Since the internal moments, like the external

138

load, increase gradually from zero to their final value as the beam is loaded, the work performed on the element by these moments is

$$dW_{\text{INT}} = \frac{M \, d\theta}{2}$$

or in view of Eq. (6.21)

$$dW_{\text{INT}} = \frac{M^2 \, dx}{2EI}$$

The total internal work for the entire beam is obtained by summing the work for all the elements. Thus

$$W_{\text{INT}} = \int_0^L \frac{M^2 \, dx}{2EI} \tag{7.7}$$

In accordance with the principle of conservation of energy we can now equate the expressions in Eqs. (7.6) and (7.7). Thus

$$\frac{P\delta}{2} = \int_0^L \frac{M^2 \, dx}{2EI} \tag{7.8}$$

Since $M = Px/2$ for $0 < x < L/2$ and since M is symmetric about the middle of the beam, Eq. (7.8) becomes

$$\frac{P\delta}{2} = 2 \int_0^{L/2} \left(\frac{Px}{2}\right)^2 \frac{dx}{2EI}$$

from which

$$\delta = \frac{PL^3}{48EI}$$

Although the method of real work led to a very simple and straightforward solution in the foregoing example, it has some serious shortcomings. The method can be applied only to a structure with a single, concentrated applied load. Furthermore, it can then be used only to calculate the deflection under that load. If one attempts to determine the deflection of some other point, not under the load, the deflection would not appear in the expression for the external work and consequently could not be evaluated. On the other hand, the presence of more than one external load would lead to an external work expression containing several unknown deflections, one for each load. However, there would still be only a single equation available to solve for these deflections.

The method of real work is thus not a very useful procedure. However, it does serve the purpose of providing a good introduction to the method of virtual work to be considered next.

7.4 VIRTUAL WORK

In this section we will consider an energy method for determining deflections that is not subject to the serious shortcomings of the procedure outlined in the previous article. In fact, of all the methods for calculating deflections that we have considered so far, the method of virtual work is the most versatile one. Whereas each of the procedures studied previously is applicable to some types of structures but not to others, the principle of virtual work is well suited for dealing with all types of structures including beams, frames, and trusses.

To help us in the development of the method, let us consider a specific problem, namely, the calculation of the deflection δ_A for the beam in Fig. 7.3a. A typical element of the beam, showing the internal moment M that acts on any cross section and the rotation $d\theta$ between two adjacent cross sections, is shown in Fig. 7.3b.

Let us now consider a second beam, shown in Fig. 7.3c. To this beam we apply,

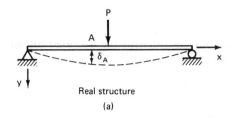

Real structure

(a)

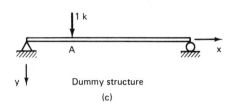

$d\theta = \dfrac{M}{EI}\,dx$

(b)

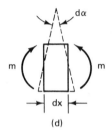

Dummy structure

(c)

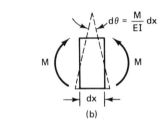

$d\alpha$

(d)

FIGURE 7.3

instead of the actual load, a unit load at the point where we are looking for the deflection and in the direction of the desired deflection. In other words, we apply a unit vertical load at A. This second beam is referred to as the *dummy structure*. To differentiate between the moments and rotations in the dummy beam and those in the real beam, we designate the moments and rotations in the dummy structure by m and α as shown in Fig. 7.3d.

We will now subject the dummy beam, on which the 1 k load is already acting, to a virtual displacement and then apply the law of conservation of energy to the resulting work. The term "virtual displacement" refers to a displacement that is not actually occurring but is only imagined to take place. Unlike a real displacement, a virtual displacement causes no change in either the external or internal forces acting on the structure.

Thus we imagine the dummy beam with the 1 k load already in place to be subjected to the deformation that the actual load P produces in the real beam. In other words, every point in the dummy beam is made to go through the deformation that occurs at that point in the real beam. As a consequence, point A moves down an amount δ_A and the 1 k load does an amount of work $1 \cdot \delta_A$. Since the 1 k force is constant throughout the displacement, there is no $\frac{1}{2}$ factor in the work term. We refer to the term $1 \cdot \delta_A$ as the *external virtual work*.

The application of the virtual displacement to the dummy beam also gives rise to internal virtual work. For each differential element the existing moment m moves through a rotation $d\theta$. If these increments of work are summed up over the entire beam, we obtain for the total *internal virtual work*

$$\text{Internal Virtual Work} = \int_0^L m \, d\theta = \int_0^L \frac{Mm \, dx}{EI}$$

Again, there is no $\frac{1}{2}$ factor present because the moments m remain constant throughout the rotations $d\theta$ produced by the virtual displacement.

We now proceed in accordance with the law of conservation of energy to equate the external virtual work to the internal virtual work. Thus

$$1 \cdot \delta_A = \int_0^L \frac{Mm \, dx}{EI} \tag{7.9}$$

Since all quantities appearing in this expression except δ_A are known, the expression can be used to evaluate δ_A.

The principle of virtual work, embodied in Eq. (7.9), differs from the method of real work in one very important way. In real work, work is the result of forces acting through displacement that they themselves cause. By comparison, in virtual work, work results when one force system acts through displacements caused by another force system. In the preceding derivation, virtual work results when the 1 k load and internal moments m of the dummy structure are made to act through the external and internal deformations produced by the actual load P on the real beam.

7.5 APPLICATION OF VIRTUAL WORK TO BEAMS AND FRAMES

The method of virtual work requires that one construct a dummy structure and apply to it a unit load at the point and in the direction of the desired deflection. The moments caused by the real loads as well as those due to the unit load are then calculated and substituted in Eq. (7.9). The solution of this equation gives the desired deflection.

Example 7.1

Determine the vertical deflection at the free end of the cantilever beam in Fig. 7.4a.

To begin, we place a unit load at point A on the dummy beam as shown in Fig. 7.4b. Next, the moments M and m needed to solve Eq. (7.9) are calculated and listed in tabular form. To carry out the integration indicated in Eq. (7.9) requires that the beam be subdivided into a sufficient number of sections so that both M and m are continuous functions of x in each section. Thus we have divided the beam into two sections, one from A to B and another from B to C. The origin for the moment expression in any section should be chosen so that the resulting expression is as simple as possible.

The moments listed in Table 7.1 are written employing point A as the origin for both segments AB and BC. Substituting these expressions into Eq. (7.9) and using the appropriate limits of integration, we obtain

$$\delta_A = \int_0^{L/2} \frac{0(x)\,dx}{EI} + \int_{L/2}^{L} \frac{P\left(x - \frac{L}{2}\right)(x)\,dx}{EI}$$

$$\delta_A = \frac{5PL^3}{48EI}$$

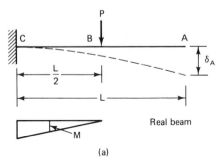

Real beam

(a)

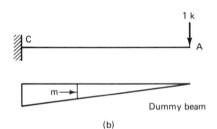

Dummy beam

(b)

FIGURE 7.4

TABLE 7.1

Element	x = 0	M	m
AB	A	0	x
BC	A	$P(x - \frac{L}{2})$	x

Note: Moments are assumed positive when they produce compression on the lower side of the beam.

TABLE 7.2

Element	x = 0	M	m
AB	A	0	x
BC	B	Px	$\frac{L}{2} + x$

Note: Moments are assumed positive when they produce compression on the lower side of the beam.

For comparison, the solution is repeated using the moments in Table 7.2. In this instance the origin for segment AB is taken at A and the origin for BC at B. As before, the limits of integration in Eq. (7.9) are chosen in accordance with the location of the origins. Thus

$$\delta_A = \int_0^{L/2} \frac{0(x)\,dx}{EI} + \int_0^{L/2} \frac{Px\left(\frac{L}{2} + x\right) dx}{EI}$$

$$\delta_A = \frac{5PL^3}{48EI}$$

Since use of the method of virtual work will not be restricted to simple beams, whose deflected shape can be obtained by inspection, it is necessary to adopt a sign convention. The following convention will be employed:

1. The same criterion must be used to define positive bending for both the real and the dummy structure; that is, both M and m must be assumed positive when they produce compression on the same side of the member.
2. A positive answer for the deflection indicates that the direction of the deflection is the same as that of the dummy load. A negative answer means that the deflection is opposite in direction to the dummy load.

For example, in the foregoing calculations both M and m were assumed to be positive when they produced compression along the lower edge of the member. Furthermore, the solution was positive, indicating that A moves down, in the direction assumed for the dummy load.

The preceding sign convention is relatively easy to comprehend. Whenever the unit load causes the dummy structure to deflect in the same direction as the actual structure deflects, m and M have the same sign and their product is positive. Conversely, if the unit load is assumed so that the dummy structure deflects in a direction opposite to that of the real structure, m and M will have opposite signs and their product will be negative.

Example 7.2

Determine the vertical deflection at C for the beam in Fig. 7.5a.

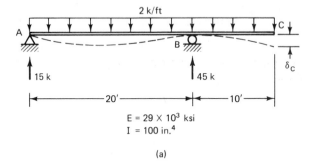

(a)

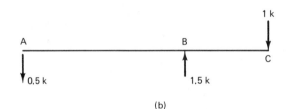

(b)

FIGURE 7.5

TABLE 7.3

Element	$x = 0$	M	m
AB	A	$-15x + x^2$	$0.5x$
BC	C	x^2	x

Note: Moments are considered positive
when they produce compression on
the lower side of the beam

To calculate the required deflection, we place a unit load on the dummy beam at C, as shown in Fig. 7.5b. Expressions for M and m are then obtained and listed in Table 7.3. The moments for segment AB are written using an origin at A and the moments for segment BC using an origin at C. The moments are assumed to be positive when they produce compression on the lower edge of the beam. Substitution of these moments into Eq. (7.9) gives

$$1 \cdot \delta_C = \int_0^{20} \frac{(-15x + x^2)(0.5x)\, dx}{EI} + \int_0^{10} \frac{(x^2)(x)\, dx}{EI}$$

$$1 \cdot \delta_C = \frac{2{,}500 \text{ kip}^2\text{-ft}^3}{EI}$$

Finally, we substitute values for E and I and make use of a conversion factor to obtain consistent dimensions. Thus

$$\delta_C = \frac{2500(1728)}{29 \times 10^3(100)} = 1.49 \text{ in}$$

The positive sign of the result indicates that δ_C is in the same direction as the dummy load, i.e., δ_C is downward.

Example 7.3

Determine both the vertical and the horizontal deflection at A for the frame in Fig. 7.6a.

To calculate the vertical deflection, we apply a unit vertical load at A to the dummy structure as indicated in Fig. 7.6b. A second dummy structure with a unit

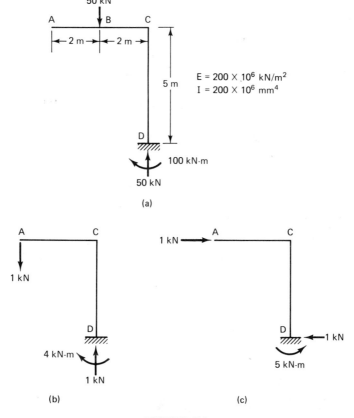

(a)

(b)   (c)

FIGURE 7.6

horizontal load at A, as shown in Fig. 7.6c, is required to calculate the horizontal deflection.

The expressions for M, the moment in the real structure, m_V, the moment in the dummy structure subjected to a unit vertical load, and m_H, the moment in the dummy structure subjected to a unit horizontal load, are listed in Table 7.4. All moments are assumed to be positive if they produce compression on the inside of the frame as indicated below the table. It does not matter how positive bending is defined. The only important factor is that the same convention is used for all moments within a given structure segment.

TABLE 7.4

Element	$x = 0$	M	m_V	m_H
AB	A	0	x	0
BC	B	$50x$	$2 + x$	0
CD	C	100	4	$-x$

Note: Moments are considered positive
when they produce compression on
the inside of the frame

Substitution of the moment expressions into Eq. (7.9) gives

$$1 \cdot \delta_V = \int_0^2 \frac{(0)(x)\, dx}{EI} + \int_0^2 \frac{(50x)(2 + x)\, dx}{EI} + \int_0^5 \frac{(100)(4)\, dx}{EI}$$

$$1 \cdot \delta_V = \frac{2333 \text{ kN}^2 \cdot \text{m}^3}{EI}$$

$$1 \cdot \delta_H = \int_0^2 0 + \int_0^2 \frac{(50x)(0)\, dx}{EI} + \int_0^5 \frac{(100)(-x)\, dx}{EI}$$

$$1 \cdot \delta_H = \frac{-1250 \text{ kN}^2 \cdot \text{m}^3}{EI}$$

from which

$$\delta_V = \frac{2333 \text{ kN} \cdot \text{m}^3}{40{,}000 \text{ kN} \cdot \text{m}^2} = 0.058 \text{ m} = 5.8 \text{ cm}$$

$$\delta_H = \frac{-1250 \text{ kN} \cdot \text{m}^3}{40{,}000 \text{ kN} \cdot \text{m}^2} = -0.031 \text{ m} = -3.1 \text{ cm}$$

The positive sign of δ_V and the negative sign of δ_H indicate that point A moves downward and to the left.

Example 7.4

Determine the rotation of joint C for the frame in Fig. 7.7a.

The method of virtual work can be used to calculate rotations or changes in slope as well as deflections. To calculate the rotation of a point in a structure, one

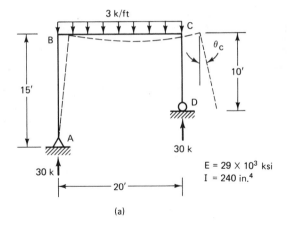

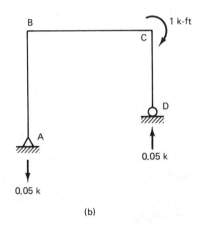

FIGURE 7.7

applies a unit couple to that point in the dummy structure. Thus we apply a unit couple to point C of the dummy structure as shown in Fig. 7.7b. Next, with the unit couple in place, we apply to the dummy structure the deformation of the actual structure as a virtual displacement. As a consequence, the unit couple moves through a rotation θ_C doing an amount of external work equal to $1 \cdot \theta_C$. At the same time the internal moments m acting on the elements of the dummy structure move through rotations $(M/EI)\,dx$ and perform internal work equal to $(m)(M/EI)\,dx$. Equating the external work to the internal work for the entire structure, we obtain

$$1 \cdot \theta_C = \int \frac{Mm\,dx}{EI} \tag{7.10}$$

This equation is very similar to Eq. (7.9). In fact, if one refers to both couples and forces as generalized forces and to deflections and rotations as generalized deflections, as is customary in mechanics, then the two equations are essentially identical.

TABLE 7.5

Element	x = 0	M	m
AB	A	0	0
BC	B	$30x - 1.5x^2$	$-0.05x$
CD	D	0	0

Note: Moments are considered positive
when they produce compression on
the outside of the frame.

The moment expressions needed to solve Eq. (7.10) are listed in Table 7.5. Moments are assumed to be positive if they produce compression on the outside of the frame. Substitution of these expressions into Eq. (7.10) gives

$$1 \cdot \theta_C = \int_0^{20} \frac{(30x - 1.5x^2)(-0.05x)\, dx}{EI} \text{ kip}^2\text{-ft}^3$$

$$1 \cdot \theta_C = -\frac{1000(144)}{(29 \times 10^3)(240)} = -0.021 \text{ radians}$$

The negative sign of the answer indicates that the joint at C rotates opposite in direction to the unit couple, i.e., θ_C is counterclockwise.

Example 7.5

Determine the vertical deflection at A for the structure in Fig. 7.8a.

To determine the desired deflection, we place a unit load on the dummy structure at A as shown in Fig. 7.8b. The internal work for this structure includes work due both to torsion and to bending. In other words, the internal work is due to torques t, of the dummy structure, rotating through angles $d\phi = (T dx)/(GJ)$ of the real structure in addition to the usual flexural work. Thus

$$1 \cdot \delta_A = \int \frac{mM\, dx}{EI} + \int \frac{tT\, dx}{GJ} \tag{7.11}$$

where T is the torque at any section in the real structure and t is the torque at the corresponding section in the dummy structure.

Substitution of the expressions for the moments and torques listed in Table 7.6 into Eq. (7.11) gives

$$1 \cdot \delta_A = \int_0^5 \frac{10x^2\, dx}{EI} + \int_0^5 \frac{15x^2\, dx}{EI} + \int_0^5 \frac{250\, dx}{GJ}$$

$$\delta_A = \frac{1042(1728)}{30 \times 10^3(144)} + \frac{1250(1728)}{12 \times 10^3(288)} = 1.04 \text{ in}$$

148

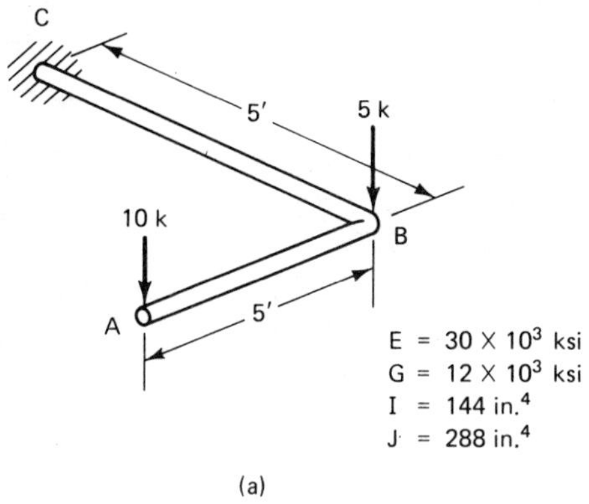
C
5′
5 k
10 k
B
A
5′
E = 30 × 10³ ksi
G = 12 × 10³ ksi
I = 144 in.⁴
J = 288 in.⁴
(a)

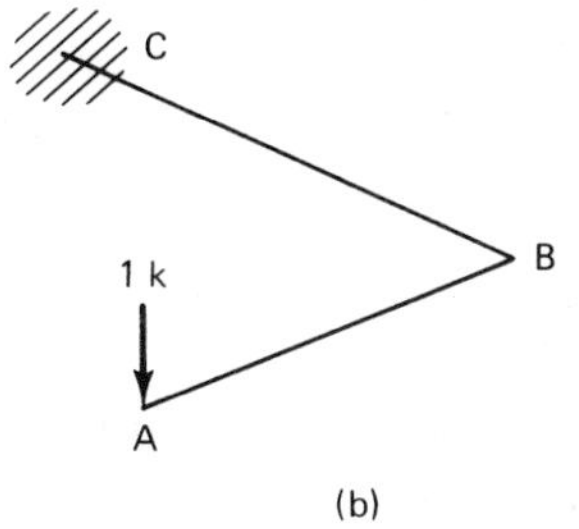
C
1 k
B
A
(b)

FIGURE 7.8

TABLE 7.6

Element	$x = 0$	M	m	T	t
AB	A	$10x$	x	0	0
BC	B	$15x$	x	50	5

7.6 DEFLECTION OF TRUSSES USING VIRTUAL WORK

One of the advantages of the method of virtual work is that it can be employed to calculate truss deflections as well as beam deflections. To aid us in the development of the basic equation needed for determining truss deflections, we will consider the truss in Fig. 7.9a. It is desired to calculate the vertical deflection of point g along the lower chord of the truss. Accordingly, we apply a unit vertical load to this point of the dummy structure as indicated in Fig. 7.9b. We then subject the dummy structure, with the 1 k load already in place, to the deformations of the real truss and equate the resulting external and internal work. The 1 k load is the only external load that performs work as the dummy structure deforms, and since this force remains constant at its full value as point g moves down a distance Δ, the external work is $1 \cdot \Delta$. To

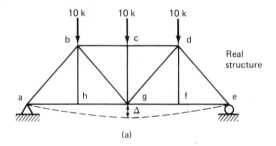

(a)

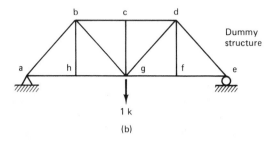

(b)

FIGURE 7.9

determine the internal work, we must first define the internal forces that are present in both the real and the dummy structure. Thus we let S represent the bar forces in the real truss and u the bar forces in the dummy structure. If the length, area, and modulus of elasticity of a bar in the truss are designated by L, A, and E, then the axial deformation of a bar in the real truss is $(SL)/(AE)$. Applying this deformation to the bars of the dummy truss in which a force u is present, we perform an amount of internal work equal to $u(SL)/(AE)$ for each bar. Summing this internal work for all the bars in the truss and equating it to the external work gives

$$1 \cdot \Delta = \sum \frac{uSL}{AE} \tag{7.12}$$

To calculate the deflection of a point in a truss, one proceeds as follows: (1) Determine the bar forces S in the actual truss; (2) apply a unit load to the dummy structure in the direction of the desired deflection and calculate the resulting bar forces u; (3) the deflection is then obtained by summing the terms $(uSL)/(AE)$ for all the bars in the truss as indicated in Eq. (7.12).

Example 7.6

Determine the vertical deflection of point g for the truss in Fig. 7.10a.

The solution of Eq. (7.12) is best carried out by making use of a table, as indicated in the figure. First the bar forces S in the real structure and the bar forces u in the dummy structure are calculated and entered in the table. Next the quantities uSL/A are determined for each bar and their sum obtained. Since E is constant, it is easier to obtain the sum of all the uSL/A terms first, and then divide by E, than to divide each

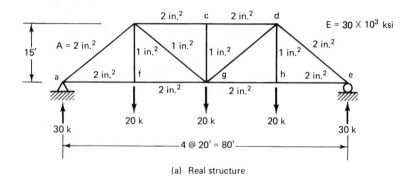

(a) Real structure

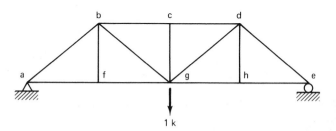

(b) Dummy structure

Member	A in.^{2}	L ft	S(k)	u(k)	uSL/A	n	nuSL/A
ab	2	25	−50	−0.83	518.75	2	1037.5
af	2	20	40	0.67	268.0	2	536.0
fg	2	20	40	0.67	268.0	2	536.0
bf	1	15	20	0	0	2	0
bg	1	25	16.7	0.83	346.5	2	693.0
bc	2	20	−53.3	−1.33	708.9	2	1417.8
cg	1	15	0	0	0	1	0
							4220.3

$$1 . \delta = \Sigma \frac{SuL}{AE}$$

$$= \frac{4220.3(12)}{30 \times 10^3}$$

$$= 1.69 \text{ in.}$$

(c)

FIGURE 7.10

of the terms by E before their sum is determined. Because of symmetry, some bars contribute two identical uSL/A terms to the calculations.

   Since the solution is positive, the truss deflects in the same direction as the dummy load, i.e., downward.

PROBLEMS

7.1 and 7.2. Use the method of real work to calculate the indicated quantities.

7.1.   Find δ_B.   **7.2.**   Find δ_B.

 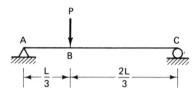

7.3 to 7.25. Use the method of virtual work to calculate the indicated quantities.

7.3.   Find δ_C.

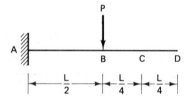

7.4.   Find δ_B.

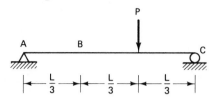

7.5.   Find δ_B; $E = 200 \times 10^6$ kN/m^2, $I = 100 \times 10^6$ mm^4.

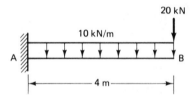

7.6. Find δ_B; $E = 30 \times 10^3$ ksi, $I = 50$ in⁴.

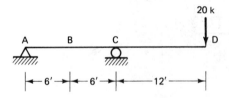

7.7. Find δ_C; $E = 200 \times 10^6$ kN/m², $I = 40 \times 10^6$ mm⁴.

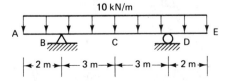

7.8. Find δ_D; $E = 200 \times 10^6$ kN/m², $I = 20 \times 10^6$ mm⁴.

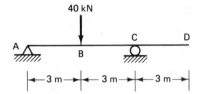

7.9. Find δ_D; $E = 30 \times 10^3$ ksi, $I = 144$ in⁴.

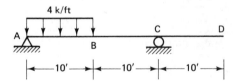

7.10. Find δ_E; $E = 30 \times 10^3$ ksi, $I = 72$ in⁴.

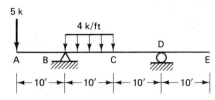

7.11. Find δ_D; $E = 30 \times 10^3$ ksi, $I = 36$ in⁴.

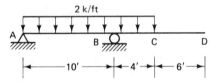

7.12. Find the vertical deflection at A; $E = 30 \times 10^3$ ksi, $I = 80$ in⁴.

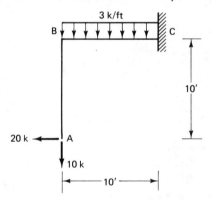

7.13. Find the horizontal deflection at D; $E = 200 \times 10^6$ kN/m², $I = 200 \times 10^6$ mm⁴.

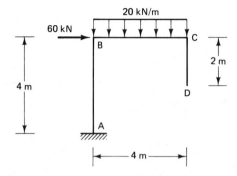

7.14. Find the horizontal deflection and the rotation at C; $E = 30 \times 10^3$ ksi, $I = 60$ in⁴.

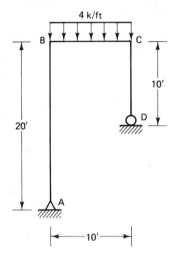

7.15. Find the vertical deflection at E; $E = 200 \times 10^6$ kN/m², $I = 400 \times 10^6$ mm⁴.

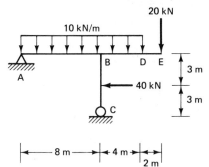

7.16. Find the rotation at B; $E = 30 \times 10^3$ ksi, $I = 72$ in⁴.

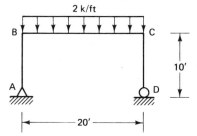

7.17. Find the rotation and the vertical deflection at A; $E = 200 \times 10^6$ kN/m², $I = 120 \times 10^6$ mm⁴.

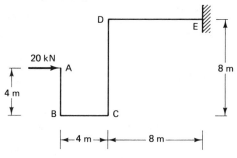

7.18. Find the horizontal deflection at E; $E = 30 \times 10^3$ ksi, $I = 576$ in⁴.

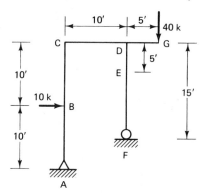

7.19. Find the vertical deflection at C; $E = 200 \times 10^6$ kN/m², $G = 80 \times 10^6$ kN/m², $I = 100 \times 10^6$ mm⁴, $J = 180 \times 10^6$ mm⁴.

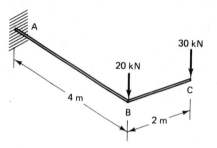

7.20. Find the vertical deflection at D; $E = 30 \times 10^3$ ksi, $G = 12 \times 10^3$ ksi, $I = 24$ in⁴, $J = 48$ in⁴.

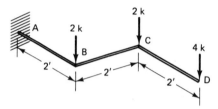

7.21. Find the vertical deflection at E; $E = 30 \times 10^3$ ksi; the area of each bar is noted in the figure.

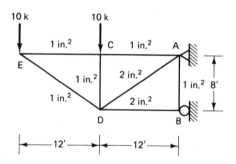

7.22. Find the vertical deflection at C; $E = 200 \times 10^6$ kN/m²; the area of each bar is equal to 500 mm².

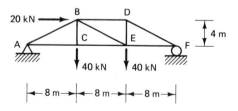

7.23. Find the vertical deflection at A; $E = 30 \times 10^3$ ksi; the area of each bar is noted in the figure.

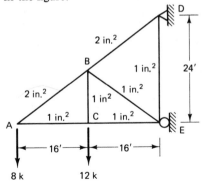

7.24. Find the deflection at C; $E = 200 \times 10^6$ kN/m², I of beam $= 100 \times 10^6$ mm⁴, A of cable $= 200$ mm².

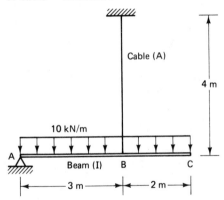

7.25. Find the deflection at B; $E = 30 \times 10^3$ ksi, I of beam $= 150$ in⁴, A of each cable $= 0.2$ in².

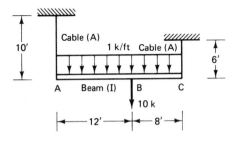

8

Influence Lines

Astoria Bridge over Columbia River, Astoria, Oreg. (*Courtesy of Oregon Department of Transportation.*)

8.1 INTRODUCTION

Many structures must resist moving loads, in addition to loads such as their own weight that remain fixed in place. For example, the force exerted by a truck on a bridge may act anywhere along the span of the bridge. Since stresses caused by moving loads will vary with the position of the load, and since structures must be designed for the largest stresses that will occur, it is necessary to determine the position of the load that produces the maximum stresses in the structure. A convenient way of doing this is to make use of influence lines.

8.2 INFLUENCE LINES DEFINED

Influence lines provide us with a systematic procedure for determining how the force in a given part of a structure varies as the applied load moves about on the structure. For example, let us determine the value of the reaction R_B, for the simply supported beam in Fig. 8.1a, due to a unit load placed at different distances x from the left-hand end of the member. Writing an equation of moment equilibrium about A gives

$$R_B = \frac{1(x)}{L} \tag{8.1}$$

It is evident from this equation that $R_B = 0$ when the load is acting directly over the left support and that $R_B = 1$ when the load is at the extreme right end of the member. Similarly, Eq. (8.1) can be used to obtain the magnitude of R_B for any intermediate position of the load. This has been done for several points, and the results are given

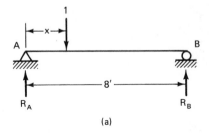

(a)

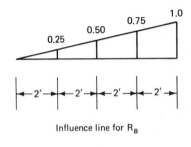

Influence line for R_B

(b)

X	R_B
0	0
2	0.25
4	0.50
6	0.75
8	1.0

FIGURE 8.1

both in tabular form and graphically in Fig. 8.1b. The diagram in the figure is referred to as an *influence line*. Its ordinate, corresponding to any value of x, gives the magnitude of R_B due to a unit load acting at x. An influence line can thus be said to depict graphically the effect, on some structural component, of a unit load as it moves along the structure. In this case, the influence line indicates that R_B varies linearly from 0 to 1 as a unit load moves from the left end to the right end of the member, and that R_B attains its maximum value when the load is directly over the right reaction.

Using the procedure defined above, influence lines can be drawn for a variety of internal and external forces such as shears and moments at different points in a beam as well as forces in various bars of a truss.

In dealing with influence lines it is important not to confuse them with moment diagrams. Whereas a moment diagram gives the value of the bending moment at different points in a beam for a load that is fixed in place, an influence line gives the value of the bending moment at one point in the beam for different locations of the load.

We will divide our study of influence lines into two parts. First we will concentrate on the procedure used to construct influence lines, and then we will make use of influence lines to determine the placement of loads for producing the maximum forces in various parts of a structure.

8.3 INFLUENCE LINES FOR BEAMS

One could obtain the influence line for a given function, such as the moment at the center of a beam, by successively placing a unit load at a series of equally spaced points along the member and evaluating the midspan moment for each position of the load. However, it is unnecessary to carry out this rather tedious procedure. Instead, influence lines will be constructed by employing the following method: A unit load is placed at a few critical points along the member, and the coordinates of the influence line corresponding to these load positions are evaluated. The sections of the influence line between these points are then constructed using basic principles of structural behavior. To see how this process is carried out, let us consider the beam in Fig. 8.2a.

The simplest way to obtain the influence line for the left reaction R_A is to write a moment equation about point B. This gives an expression for R_A that is valid for any position x of the unit load. Thus

$$R_A(L) - 1(L - x) = 0$$

$$R_A = 1 - \frac{x}{L} \tag{8.2}$$

Equation (8.2) indicates that R_A varies linearly from 1 to 0 as x varies from 0 to L. This result is depicted graphically by the influence line in Fig. 8.2b. In drawing the influence line we have assumed upward reactions to be positive.

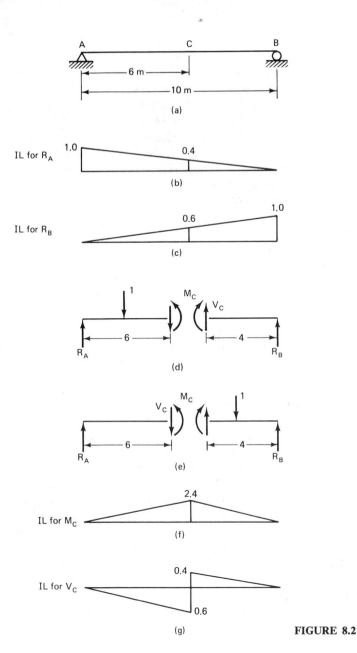

FIGURE 8.2

Similarly, writing a moment equation about A gives

$$R_B = \frac{x}{L}$$

which leads to the influence line in Fig. 8.2c.

The foregoing results lead to the following conclusions regarding influence lines for reactions of simply supported beams: (1) When the load is directly over one of the supports, the reaction at that support resists the entire load. (2) As the load moves away from the support, the value of the reaction at the support decreases linearly until the load reaches the other support, at which time the value of the reaction at the first support is zero. (3) From vertical equilibrium, the ordinates of the influence lines for the two reactions must add up to unity for any point on the beam.

Let us now draw the influence line for the bending moment at C. To obtain an expression for M_C, it is necessary to consider a free body of the beam either to the right or to the left of C, as shown in Figs. 8.2d and 8.2e. When the unit load is to the left of C (Fig. 8.2d), the right-hand free body is the simpler of the two to use for determining M_C. From this free body

$$M_C = 4R_B, \, 0 \leq x \leq 6$$

According to this relation, the segment of the influence line for M_C between A and C, shown in Fig. 8.2f, is obtained by taking the influence line for R_B between A and C and multiplying it by 4. The influence line has been drawn with the assumption that positive bending corresponds to compression on the upper part of the beam.

To obtain the influence line for M_C when the load is between C and B, we consider the free body to the left of C in Fig. 8.2e. From this free body,

$$M_C = 6R_A, \quad 6 \leq x \leq 10$$

Thus the section of the influence line for M_C between C and B is obtained by taking the influence line for R_A between C and B and multiplying it by 6.

The influence line for the shear at C is drawn using a procedure similar to that employed for constructing the moment influence line. The right-hand free body in Fig. 8.2d is used to obtain the influence line for V_C when the load is between A and C, leading to

$$V_C = -R_B, \quad 0 \leq x \leq 6$$

and the left-hand free body in Fig. 8.2e is used when the load is between C and B, giving

$$V_C = R_A, \quad 6 \leq x \leq 10$$

The above relations indicate that V_C is equal and opposite in sign to R_B when the load is between A and C and that V_C is equal to R_A when the load is between C and B. The influence line for V_C, shown in Fig. 8.2g, is thus obtained directly from the influence lines for R_B and R_A.

It should be noted that we assumed a positive shear to be a force causing a clockwise moment about any point in the free body on which the force is acting.

Example 8.1

For the beam in Fig. 8.3a, construct influence lines for the following: the reactions at A and B, the moments at A and D, and the shears at D and just to the left of A.

To begin, we place a unit load at C and determine the reactions at A and B (Fig. 8.3b). This gives us the ordinates of the influence lines for R_A and R_B at point C. Next we move the unit load from C to A (Fig. 8.3c). The effect of this on R_B can be determined by considering moment equilibrium about A. The moment of the unit load, which varies from 10 to 0 as the load moves from C to A, must be balanced by

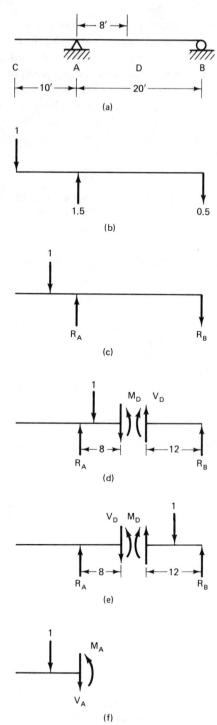

FIGURE 8.3

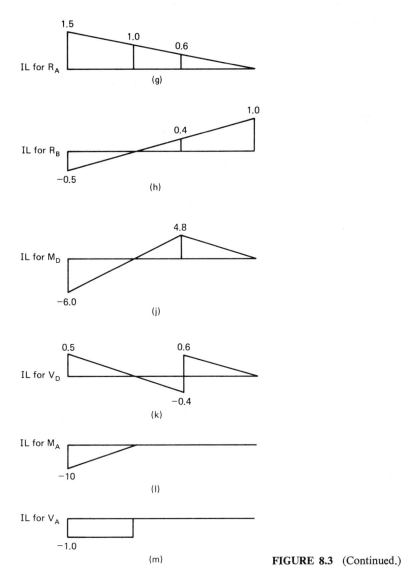

IL for R$_A$ (g)

IL for R$_B$ (h)

IL for M$_D$ (j)

IL for V$_D$ (k)

IL for M$_A$ (l)

IL for V$_A$ (m)

FIGURE 8.3 (Continued.)

$20R_B$. Thus the magnitude of R_B decreases from 0.5 to 0 (Fig. 8.3h). Similarly, it is evident from vertical equilibrium that $R_A = 1 + R_B$ and that R_A therefore decreases from 1.5 to 1 as R_B decreases from 0.5 to 0 (Fig. 8.3g).

Once the load is to the right of R_A, the overhang no longer has any effect on the behavior of the beam. The structure now behaves as if it were a simply supported beam. Thus R_A varies from 1 to 0 and R_B from 0 to 1, as the unit load moves from A to B.

Having obtained the influence lines for R_A and R_B, we can construct the influence lines for M_D and V_D by writing equations which relate M_D and V_D to R_A and R_B. This is accomplished by using Figs. 8.3d and 8.3e. When the load is between C and

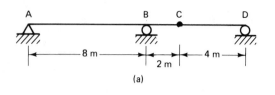

(a)

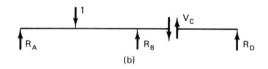

(b)

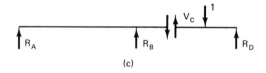

(c)

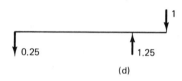

(d)

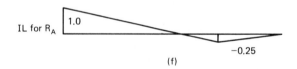

IL for R_A

(f)

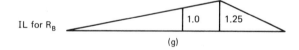

IL for R_B

(g)

IL for R_D

(h)

IL for V_C

(j)

IL for M_B

(k)

FIGURE 8.4

D, we use the right-hand free body in Fig. 8.3d, and when the load is between D and B we employ the left-hand free body in Fig. 8.3e.

The influence lines for M_A and for V_A just to the left of A are obtained by making use of the free body in Fig. 8.3f. If the unit load is on the free body, $V_A = 1$ and M_A is equal to the distance between the unit load and A. When the load is between A and B, it is evident from a free body of the overhang that both V_A and M_A must be equal to zero.

Example 8.2

For the beam in Fig. 8.4a, construct influence lines for the reactions at A, B, and D, the moment at B, and the shear at C.

To draw the influence lines for the reactions and V_C, we make use of the free bodies to the right and to the left of the hinge shown in Figs. 8.4b and 8.4c. In both figures the left-hand free body has three unknown forces acting on it. Consequently it is necessary, regardless of the position of the load, to analyze the right-hand free body first.

Let us start by considering the case depicted in Fig. 8.4b, namely, the unit load located to the left of the hinge. For this condition moment equilibrium of the right-hand free body indicates that $R_D = V_C = 0$. Once V_C is known to be zero, the left-hand free body becomes a determinate structure that can be used to evaluate R_A and R_B. With the load between A and B, segment AB behaves like a simply supported beam; that is, as the load moves from A to B, R_A varies from 1 to 0 and R_B from 0 to 1. When the load is just to the left of the hinge, R_A and R_B take on the values shown in Fig. 8.4d.

Next we consider the case where the load is to the right of the hinge as shown in Fig. 8.4c. From the right-hand free body in this figure it is evident that V_C varies from 1 to 0 and R_D from 0 to 1 as the unit load moves from C to D. Furthermore, the free body in Fig. 8.4d indicates that R_A and R_B vary from 0.25 to 0 and from 1.25 to 0, respectively, as V_C decreases from 1 to 0.

To construct the influence line for M_B, we proceed as follows. When the unit load is to the left of the hinge (Fig. 8.4b), it has been demonstrated that $V_C = 0$. Consequently $M_B = 0$ when the load is between A and B, and M_B is equal to the unit load multiplied by its distance to B when the load is between B and C. Finally, when the load is to the right of the hinge (Fig. 8.4c), the magnitude of M_B is two times that of V_C.

8.4 INFLUENCE LINES FOR TRUSSES

Influence lines for bar forces in trusses are constructed using the same general procedure that was employed to obtain influence lines for shears and moments in beams. A unit load is moved along the loaded chord of the truss, and the resulting value of the bar force is plotted versus the location of the load. One difference between beams and trusses is that the load can be applied anywhere along the span of a beam, whereas loads are usually applied to trusses only at their joints. As shown in Fig. 8.5, the steel or concrete deck of a truss bridge rests on longitudinal members called *stringers*. These stringers, in turn, span between transverse members called *floor beams*, and the floor beams frame into the joints of the trusses. Loads applied to the deck of the

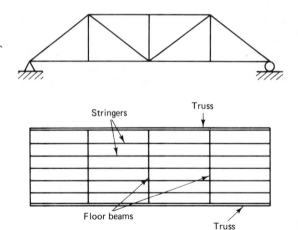

Stringers

Truss

Floor beams

Truss

FIGURE 8.5

bridge are thus transferred from the deck to the stringers, from the stringers to the floor beams, and thence to the joints of the trusses. As a result, a load applied anywhere along the deck of the bridge is felt by the trusses only at their joints.

To see how influence lines for bar forces in a truss are constructed, let us consider the truss in Fig. 8.6a. The load is assumed to be applied to the truss along the lower chord.

We begin by drawing the influence lines for the reactions at A and G. These influence lines, shown in Figs. 8.6c and 8.6d, are identical to those for a beam. Next we construct the influence line for the force in member BC. We can obtain an expression for F_{BC} by considering the free body of the truss either to the right or to the left of a vertical section between joints B and C (Fig. 8.6b).

As long as the unit load is to the left of J, the right-hand free body is the easier of the two to use. Taking moments about J gives

$$-15F_{BC} - 80R_G = 0$$

or

$$F_{BC} = -5.33R_G$$

The negative sign indicates that F_{BC} is compression when R_G is positive. Thus the segment of the influence line for F_{BC} between A and J, shown in Fig. 8.6e, is obtained by multiplying the corresponding influence line segment for R_G by -5.33.

When the unit load is to the right of J, the left-hand free body in Fig. 8.6b is used. Summing moments about J of the forces acting on the free body leads to

$$15F_{BC} + 40R_A = 0$$

or

$$F_{BC} = -2.67R_A$$

Using this expression, we construct the segment of the influence line for F_{BC} between J and G.

The free bodies in Fig. 8.6b can also be used to draw the influence line for F_{BJ}. If the unit load is to the left of H, we write an equation of vertical equilibrium for

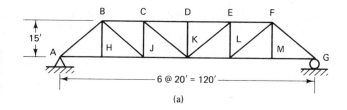

(a)

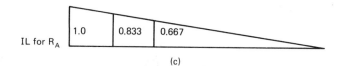

(b)

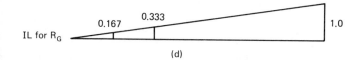

(c)

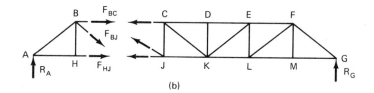

(d)

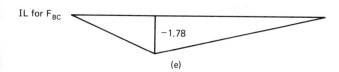

(e)

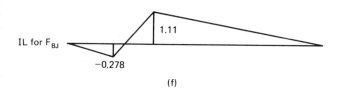

(f)

FIGURE 8.6

the right-hand free body. Thus

$$0.6F_{BJ} + R_G = 0$$

or

$$F_{BJ} = -1.67R_G$$

This relation leads to the segment of the influence line between A and H shown in Fig. 8.6f.

When the unit load is between H and J, we can continue to use the right-hand free body in Fig. 8.6b. However, a vertical downward load that varies from 0 to 1 as the unit load moves from H to J is now applied at J in addition to the forces shown in Fig. 8.6b. If the unit load is located at a distance x from H, vertical equilibrium of the free body gives

$$0.6F_{BJ} + R_G - \frac{x}{20} = 0$$

or

$$F_{BJ} = -1.67R_G + \frac{x}{12}$$

The force F_{BJ} thus varies linearly from $-1.67(0.167) = -0.278$ when the unit load is at H, to $-1.67(0.333) + 1.67 = 1.11$ when the unit load is at J.

Finally, when the unit load is to the right of J, we write an equation of vertical equilibrium for the left-hand free body in Fig. 8.6b and obtain

$$0.6F_{BJ} = R_A$$

or

$$F_{BJ} = 1.67R_A$$

This expression is used to construct the influence line segment for F_{BJ} between J and G.

Example 8.3

Construct influence lines for the forces in members AB, BF, and BG, for the truss in Fig. 8.7a. The loads are applied to the truss at the joints of the lower chord.

As the first step the influence lines for the reactions at A and E, shown in Figs. 8.7d and 8.7e, are drawn.

Next the influence line for F_{AB} is obtained by considering the free bodies in Fig. 8.7b. When the unit load is between A and F, F_{AB} can be determined by taking moments about F of the forces acting on the right-hand free body. Thus

$$-0.83F_{AB}(10) - 45R_E = 0$$

or

$$F_{AB} = -5.42R_E$$

This expression leads to the influence line segment for F_{AB} between A and F.

When the unit load is to the right of F, we take moments about F of the forces acting on the left-hand free body in Fig. 8.7b. This leads to

$$0.83F_{AB}(10) + 15R_A = 0$$

or

$$F_{AB} = -1.81R_A$$

from which the influence line segment between F and E is constructed.

To draw the influence line for F_{BF}, we consider a free body of joint F. This indicates that a force exists in bar BF only when a vertical load is applied to the joint, in which case F_{BF} is equal to that load. Since the load applied to joint F varies from 0 to 1 as the unit load moves from A to F and from 1 to 0 as the unit load moves from F to G, the influence line for F_{BF} varies from 0 to 1 and back to 0 between A and G, as shown in Fig. 8.7g.

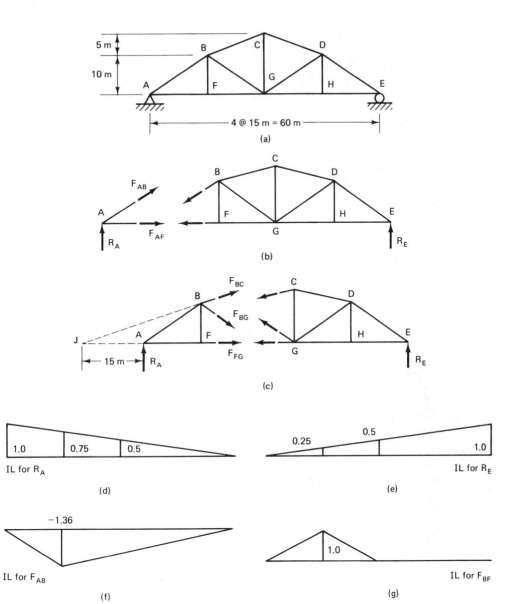

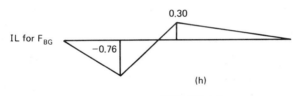

FIGURE 8.7

When the unit load is between G and E, the force in bar BF is zero.

Finally, to construct the influence line for F_{BG} we make use of the free bodies in Fig. 8.7c. When the unit load is between A and F, we take moments about J of the forces acting on the right-hand free body. This gives

$$-0.55F_{BG}(45) - R_E(75) = 0$$

or

$$F_{BG} = -3.03R_E$$

Similarly, when the unit load is between G and E, we take moments about J of the forces acting on the left-hand free body. Thus

$$0.55F_{BG}(45) - R_A(15) = 0$$

or

$$F_{BG} = 0.61R_A$$

Using the foregoing results, we construct the influence line segments between A and F and between G and E as shown in Fig. 8.7h. The last segment of the influence line, between F and G, is obtained by drawing a straight line between the previously constructed segments.

The validity of the last step can be checked by placing the unit load halfway between F and G and summing moments about J for the right-hand free body. This leads to

$$-0.55F_{BG}(45) - 75R_E + 0.5(45) = 0$$

or

$$F_{BG} = -0.23$$

which checks with the value at the midpoint of the influence line segment between F and G in Fig. 8.7h.

8.5 USES OF INFLUENCE LINES

Having learned how to construct influence lines we are now ready to make use of them.

Concentrated Loads

To see how influence lines can be employed to determine the effects of moving concentrated loads, let us consider the beam in Fig. 8.8a and the accompanying influence line for the moment at C. The ordinate of the influence line, corresponding to any point on the beam, gives the value of M_C when a unit load is placed at that point. Thus a 1 k load placed 6 ft to the right of A results in a moment $M_C = 2.4$ k-ft. Similarly a 15 k load placed in the same positions gives rise to a moment fifteen times as large, i.e., $M_C = 36$ k-ft.

To obtain the value of M_C due to several concentrated loads, one multiplies each load by the ordinate of the influence line corresponding to the position of the load and then adds the resulting moments. For example, the three loads, placed as

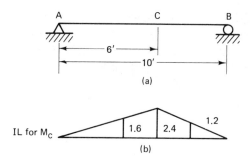

(a)

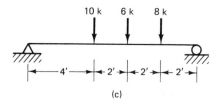

IL for M_C

(b)

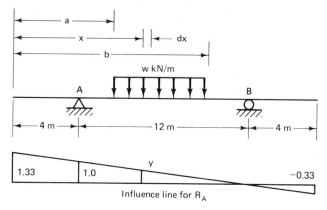

(c)

FIGURE 8.8

indicated in Fig. 8.8c, give rise to a moment

$$M_C = 10(1.6) + 6(2.4) + 8(1.2) = 40 \text{ k-ft}$$

Uniformly Distributed Loads

Although influence lines, by definition, represent the effect of concentrated loads, they can be used to deal with uniformly distributed loads as well. To see how this is accomplished, let us consider the beam and the accompanying influence line for R_A in Fig. 8.9. As indicated, the beam is loaded by a uniformly distributed load, of intensity w, extending from $x = a$ to $x = b$. Because of this loading an element dx of the beam, located a distance x from the left end, has acting on it a load equal to

FIGURE 8.9

wdx. Furthermore, if the ordinate of the influence line at x is equal to y, the value of R_A due to the load increment wdx is $(wdx)(y)$. If we now think of the entire load between $x = a$ and $x = b$ as consisting of a series of differential load increments, the value of R_A due to these increments is equal to the sum of the $(wdx)(y)$ terms for the interval from $x = a$ to $x = b$. Thus

$$R_A = \int_a^b wy \, dx$$

and since w is a constant,

$$R_A = w \int_a^b y \, dx \qquad (8.3)$$

Equation (8.3) indicates that the value of R_A due to a uniformly distributed load is equal to the intensity of the load multiplied by the area of the influence line corresponding to the length of beam over which the load is acting.

For example, a distributed load, $w = 4$ kN/m, extending from A to B, will result in a reaction at A equal to

$$R_A = (4)(1)(12)(0.5) = 24 \text{ kN}$$

If the same load acts over the entire beam, instead of simply from A to B, we obtain

$$R_A = 4(1.33)(16)(0.5) - 4(0.33)(4)(0.5) = 40 \text{ kN}$$

The influence line for R_A indicates that loads to the left of B cause an upward force R_A, whereas loads to the right of B result in a downward reaction at A. As a consequence the second quantity in the foregoing calculation is subtracted from the first one.

Example 8.4

For the beam in Fig. 8.10a, find the maximum value of M_D due to a uniform load of intensity 5 kN/m which can act over any part of the beam, and two concentrated loads of 10 kN each, with a fixed distance of 4 meters between them.

First we construct the influence lines for the reactions and from these the influence line for M_D shown in Fig. 8.10d. Next we decide where to place the loads in order to produce the largest possible value of M_D.

Since loads acting in a region where the influence line is positive have an effect that is opposite from that produced by loads acting where the influence line is negative, we must place all the loads either to the left or to the right of B.

The maximum value that M_D can attain if the loads are all acting to the left of B is obtained when the loads are placed as indicated in Fig. 8.10e. This leads to

$$M_D = 10(1.33) + 10(2.67) + 5(12)(2.67)(0.5) = +120 \text{ kN-m}$$

If the loads are to the right of B, they must be placed as shown in Fig. 8.10f. This loading produces a moment at D equal to

$$M_D = -10(4) - 10(1.33) - 5(4)(6)(0.5) = -113.3 \text{ kN-m}$$

Thus 120 kN-m is the largest moment at D that the given set of loads can produce.

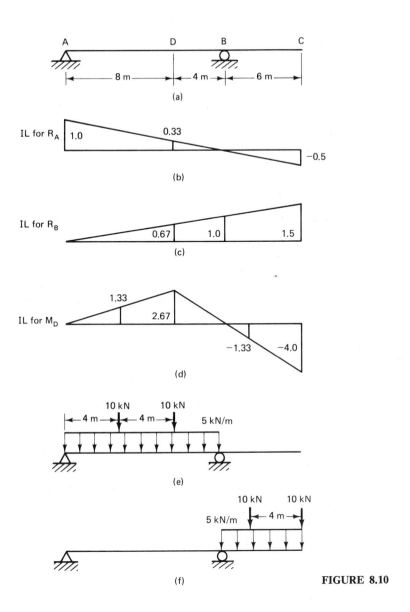

IL for R_A

1.0   0.33
              −0.5

(b)

IL for R_B

0.67   1.0   1.5

(c)

IL for M_D

1.33
        2.67
                −1.33   −4.0

(d)

10 kN   10 kN
4 m   4 m   5 kN/m

(e)

10 kN   10 kN
5 kN/m   4 m

(f)

FIGURE 8.10

PROBLEMS

8.1 to 8.15. Draw influence lines for the indicated quantities.

8.1. Reactions at A and B, moment at C.

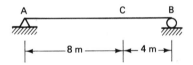

8 m   4 m

8.2.   Reactions at *A* and *B*, moment and shear at *C*.

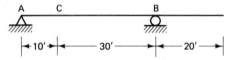

8.3.   Moment and vertical reaction at *A*, moment and shear at *B*.

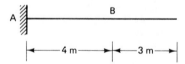

8.4.   Reactions at *A* and *B*, moment and shear at *C*.

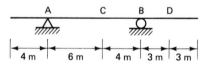

8.5.   Reactions at *A* and *B*, moment and shear at *C*, moment at *B*, shear to the left of *B*.

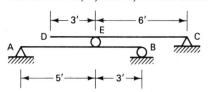

8.6.   Reactions at *A* and *B*, moments at *A*, *C*, and *D*, shear at *C* and *D*.

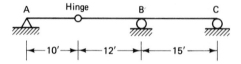

8.7.   Reactions at *A*, *B*, and *C*, as load moves from *D* to *C* along member *DEC*.

8.8.   Reactions at *A*, *B*, and *C*.

8.9.   Reactions at *A*, *B*, and *C*, moments at *B* and *D*, shear at *D* and *E*.

8.10. Vertical reactions at A and C, moment at A, shear at B.

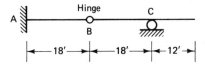

8.11. Reactions at A and B, moment at B, shear at F.

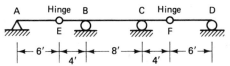

8.12. Bar forces BH, BC, BG, and GF.

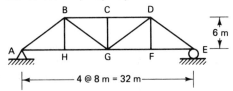

8.13. Bar forces AB, LK, BM, and CL.

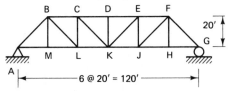

8.14. Bar forces BH, BC, BG, and DE.

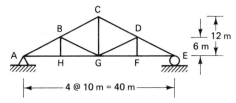

8.15. Bar forces BC, HG, BG, and DF.

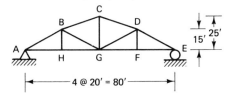

8.16. Determine the maximum value of M_C, due to two 20 k concentrated loads, that must remain 4 ft apart but can be placed anywhere along the span.

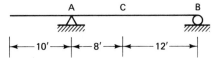

8.17.   Determine the maximum value of the reaction at *A* and the moment at *B*, due to
a uniformly distributed load of 10 kN/m, that can be placed over any portion or
portions of the span.

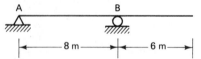

8.18.   Determine the maximum value of the reaction at *A* and the moment and shear at
C, due to a uniformly distributed load of 2 k/ft and a concentrated load of 20 k,
both of which can be placed anywhere along the span.

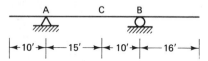

8.19.   Determine the maximum moment and shear at *C*, due to a uniform load of
12 kN/m and two 40 kN concentrated loads that must remain 2 m apart. Both the
concentrated loads and the distributed load can be placed anywhere along the span.

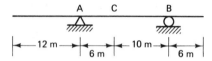

New River Gorge Bridge,
Charleston, W. Va.
*(Courtesy of American
Bridge Division, U.S. Steel
Corporation.)*

9

Arches
and Cables

9.1 ARCHES

The arch is a structural system in which the primary internal force is axial compression. Most arches do develop some bending in addition to the compression. However, the bending moments that a given set of loads produces in an arch are much smaller than those produced in a beam of the same length. As a consequence, arches are able to span much larger distances than beams.

Unlike a beam, which develops only vertical reactions at its supports, the arch gives rise to both horizontal thrusts and vertical forces at its supports. In fact, it is the horizontal reactions that are responsible for the relatively small bending moments in an arch. Without these forces the arch would simply be a curved beam whose moments would be identical to those of an equivalent straight beam. Thus the moment at midspan for both the curved and straight beams in Fig. 9.1 is equal to 500 k-ft. By comparison, the moment at the middle of the arch is only 158 k-ft.

The arch in Fig. 9.2a has two hinges, one at each support, and is referred to as a

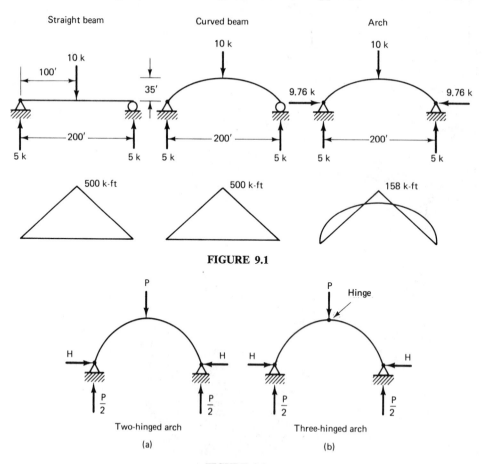

FIGURE 9.1

FIGURE 9.2

two-hinged arch. This system has more unknown reactions than equations of equilibrium and is consequently indeterminate. In a later chapter, after methods for analyzing indeterminate structures have been introduced, we will consider two hinged arches. However, for the present we will restrict our considerations to *three-hinged arches*, which are determinate. As shown in Fig. 9.2b, a three-hinged arch has a hinge at its crown in addition to those at the supports.

Example 9.1

Determine the reactions for the arch in Fig. 9.3a.

If the supports are at the same elevation, the vertical reactions are determined using the same procedure as was employed for beams. Summing moments about B for the free body in Fig. 9.3a gives

$$V_A(40) - 100(30) - 60(12) = 0$$

$$V_A = 93 \text{ kN}$$

and from vertical equilibrium

$$V_B = 160 - 93 = 67 \text{ kN}$$

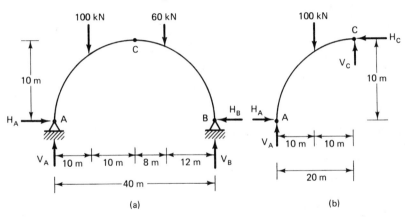

FIGURE 9.3

To determine the horizontal reaction, we sum moments about C for the free body in Fig. 9.3b. Thus

$$-H_A(10) + 93(20) - 100(10) = 0$$

$$H_A = 86 \text{ kN}$$

and from horizontal equilibrium of the entire arch we obtain

$$H_B = H_A = 86 \text{ kN}$$

It should be noted that the absence of a moment at C in Fig. 9.3b, which is due to the presence of a hinge at that point, allows us to evaluate the horizontal reactions.

Example 9.2

Determine the reactions and construct the moment diagram for the arch in Fig. 9.4.

Since the supports are not at the same elevation, the reactions cannot be deter-

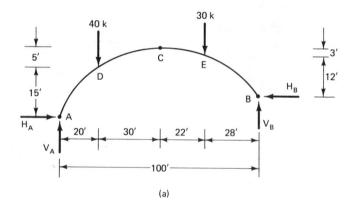

(a)

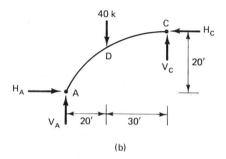

(b)

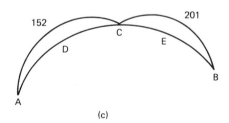

(c)

FIGURE 9.4

mined without solving simultaneous equations. Taking moments about B in Fig. 9.4a and moments about C in Fig. 9.4b gives

$$V_A(100) - H_A(5) - 40(80) - 30(28) = 0$$

and

$$V_A(50) - H_A(20) - 40(30) = 0$$

from which

$$H_A = 46.9 \text{ k}, \qquad V_A = 42.8 \text{ k}$$

The remaining two reactions are determined using equations of equilibrium for the entire arch. Thus

$$V_B = 27.2 \text{ k}, \qquad H_B = 46.9 \text{ k}$$

The moment diagram for the arch is shown in Fig. 9.4c.

9.2 CABLES

The cable is a structural element that resists forces by developing tension stresses. Because of its flexibility it is unable to develop either compression or bending stresses. When subjected to transverse loads, a cable adjusts its shape to the loads in such a manner that tension forces are sufficient to resist the loads. Concentrated loads cause the cable to take on a shape consisting of a series of linear segments, and when subjected to a distributed load the cable becomes a curve. If the load is uniformly distributed in the horizontal direction, the curve is a parabola.

Example 9.3

Determine the reactions and the tension in each segment, for the cable in Fig. 9.5a.

The reactions are determined using the same procedure as was employed in analyzing three-hinged arches. Summing moments about E for the entire system gives

$$60V_A - 12(50) + 12(25) + 8(15) = 0$$

$$V_A = 17 \text{ kN}$$

and from vertical equilibrium

$$V_E = 32 - 17 = 15 \text{ kN}$$

To solve for the horizontal reaction, we take moments about C for the free body in Fig. 9.5b. Thus

$$-H_A(12) + 17(35) - 12(25) = 0$$

$$H_A = 24.6 \text{ kN}$$

and from horizontal equilibrium

$$H_E = H_A = 24.6 \text{ kN}$$

The tension in each segment of the cable is determined using free bodies of the individual segments as shown in Fig. 9.5c. Thus

$$T_{AB} = \sqrt{(17)^2 + (24.6)^2} = 29.9 \text{ kN}$$
$$T_{BC} = \sqrt{(5)^2 + (24.6)^2} = 25.1 \text{ kN}$$
$$T_{CD} = \sqrt{(7)^2 + (24.6)^2} = 25.6 \text{ kN}$$
$$T_{DE} = \sqrt{(15)^2 + (24.6)^2} = 28.8 \text{ kN}$$

The preceding calculations indicate that the horizontal component of the cable tension is constant throughout, and that the maximum cable tension therefore occurs in the segment with the largest vertical component, i.e., the one with the largest slope.

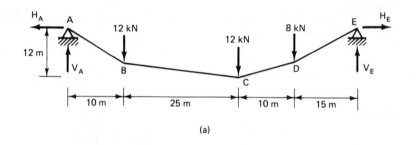

(a)

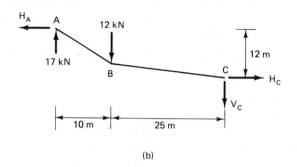

(b)

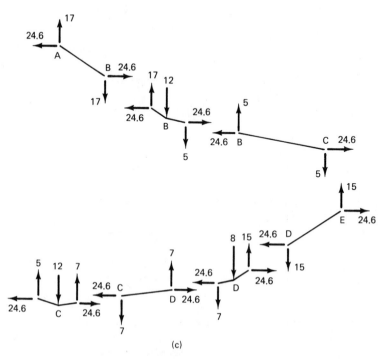

(c)

FIGURE 9.5

184

PROBLEMS

9.1 to 9.3. Determine the reactions and sketch the moment diagram.

9.1.

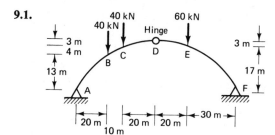

9.2.

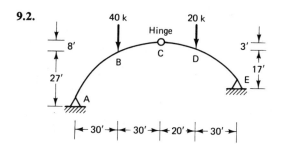

9.3.

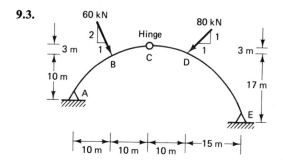

9.4 and 9.5. Determine the reactions and the indicated bar forces.

9.4. *BC, JD,* and *HG.*

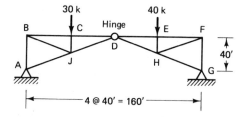

9.5. *CD*, *CK*, and *EJ*.

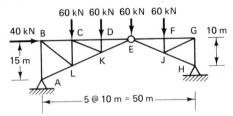

9.6 and 9.7. Determine the reactions and the tension in each segment of the cable.

9.6.

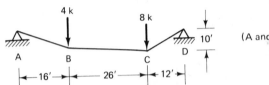

(A and D are at same level)

9.7.

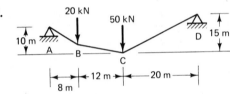

(D is 5m above A)

10

Indeterminate Structures: Introduction

Sears Tower, Chicago, Ill.
*(Courtesy of American
Bridge Division, U.S. Steel
Corporation.)*

Since determinate structures are easier to analyze than indeterminate structures, we limited ourselves to the former in the preceding chapters. However, many real structures are indeterminate, and we will therefore devote the remaining sections of the book to a study of indeterminate structural analysis. Before we proceed to consider specific methods for analyzing indeterminate structures, it is important that we form some understanding of what an indeterminate structure is and how it differs from a determinate structure.

A structure is said to be determinate if the equations of equilibrium suffice to calculate all the external reactions as well as the internal forces. By comparison, an indeterminate structure is one for which the equations of equilibrium are insufficient for determining the reactions and internal forces. To analyze an indeterminate structure, it is necessary to supplement the equations of equilibrium with additional equations obtained from a consideration of the deformation of the system. To see how these concepts apply to an actual structure, let us consider the two-bar and three-bar structures in Figs. 10.1 and 10.2. In each of these systems the bars are hinged to

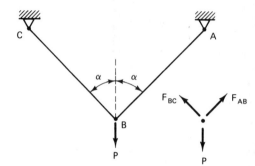

FIGURE 10.1

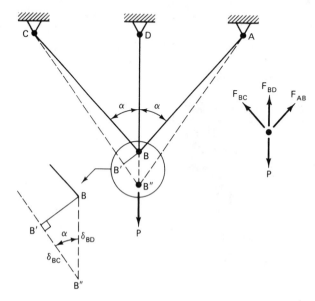

FIGURE 10.2

each other and to the supports. Thus only axial forces can be developed in the individual bars. Let us first consider the two-bar structure in Fig. 10.1. If we isolate a free body of joint B, as shown in the figure, we can write two independent equations of equilibrium. Thus

$$\sum F_x = 0$$
$$F_{AB} \sin \alpha = F_{BC} \sin \alpha$$
$$F_{AB} = F_{BC} \tag{10.1}$$

$$\sum F_y = 0$$
$$F_{AB} \cos \alpha + F_{BC} \cos \alpha = P \tag{10.2}$$

Combining Eqs. (10.1) and (10.2) gives

$$F_{AB} = F_{BC} = \frac{P}{2 \cos \alpha}$$

It is obvious from these results that the two-bar system in Fig. 10.1 is a determinate structure that can be analyzed completely using only the equations of equilibrium.

We now turn our attention to the three-bar system in Fig. 10.2. As before, we write equations of vertical and horizontal equilibrium for joint B. Thus

$$\sum F_x = 0$$
$$F_{AB} \sin \alpha = F_{BC} \sin \alpha$$
$$F_{AB} = F_{BC} \tag{10.3}$$

$$\sum F_y = 0$$
$$F_{AB} \cos \alpha + F_{BC} \cos \alpha + F_{BD} = P \tag{10.4}$$

Combining Eqs. (10.3) and (10.4) leads to

$$2F_{BC} \cos \alpha + F_{BD} = P \tag{10.5}$$

For this structure the available equations of equilibrium are not sufficient to obtain the force in each bar, and the structure is therefore said to be indeterminate. To calculate the bar forces, we must consider the deformation of the structure as well as the conditions of equilibrium.

If the bars deform as indicated by the dashed lines in the figure, then δ_{BC}, the elongation of bar BC, is related to δ_{BD}, the elongation of member BD, by

$$\delta_{BC} = \delta_{BD} \cos \alpha \tag{10.6}$$

In writing this equation we have assumed that the angle α between bars BC and BD does not change significantly as the structure deforms. This is true as long as the elongations of the bars are small compared to the original lengths of the members.

In order to solve Eqs. (10.5) and (10.6) simultaneously, it is necessary to relate the bar elongations to the bar forces. To this end, we introduce the force-deformation relation for axially loaded bars, namely,

$$\delta = \frac{Fl}{AE} \tag{10.7}$$

Since Eq. (10.7) is valid only as long as Hooke's law is satisfied, we have now introduced a second assumption, namely, that the material obeys Hooke's law. For most engineering materials this is equivalent to stating that the material remains elastic. Substitution of Eq. (10.7) into Eq. (10.6) gives

$$\frac{F_{BC}l}{A_{BC}E_{BC}} = \frac{F_{BD}l\cos^2\alpha}{A_{BD}E_{BD}}$$

or if all bars are made of the same material and have equal areas

$$F_{BC} = F_{BD}\cos^2\alpha \tag{10.8}$$

Combining this expression with Eq. (10.5), one obtains for the bar forces

$$F_{BD} = \frac{P}{1 + 2\cos^3\alpha} \tag{10.9}$$

and

$$F_{BA} = F_{BC} = \frac{P\cos^2\alpha}{1 + 2\cos^3\alpha} \tag{10.10}$$

The preceding example demonstrates clearly that whereas the equations of equilibrium suffice to determine the internal forces and reactions of a determinate structure, they do not suffice when calculating these quantities in an indeterminate structure. For the latter it is necessary to consider the deformations of the structure in addition to the equations of equilibrium.

11

Method
of Consistent Deformations

Kingston Bridge over
Hudson River, Kingston,
N.Y. (*Courtesy of
Steinman, Boynton,
Gronquist, and Birdsall.*)

11.1 INTRODUCTION

In Chapter 10, an indeterminate structure was defined as a structure in which the number of unknown forces exceeds the number of equations of equilibrium available for the calculation of these forces. For example, the beam in Fig. 11.1a is a determinate structure. It has two unknown reactions R_A and M_A, and there exist two equations of equilibrium that can be used to solve for these reactions. By comparison, the beam in Fig. 11.1b is an indeterminate structure. It possesses three unknown reactions, or one more reaction than the number of available equations of equilibrium.

For the determinate structure in Fig. 11.1a there is only one set of reactions that satisfies equilibrium, namely $R_A = P$ and $M_A = PL/2$. On the other hand, for the indeterminate beam there exist an infinite number of combinations of reactions that satisfy equilibrium. For example, $R_A = R_B = P/2$ and $M_A = 0$ satisfies equilibrium, as does $R_A = P$, $R_B = 0$ and $M_A = PL/2$. However, of the many solutions that satisfy equilibrium in an indeterminate structure only one results in a deflected shape that is compatible with the existing boundary conditions of the structure. In other words, for the structure in Fig. 11.1b, only one of the set of reactions that satisfy equilibrium will also ensure that the deflections at A and B and the slope at A are zero. The method of consistent deformations, to be presented in this chapter, makes use of this principle of deformation compatibility to analyze indeterminate structures.

All methods used to analyze indeterminate structures employ equations that relate the forces acting on the structure to the deformations of the structure. If these relations are formed so that the deformations are expressed in terms of the forces, then the forces become the independent variables or unknowns in the analysis. Methods of this type are referred to as *force methods*. On the other hand, if the basic

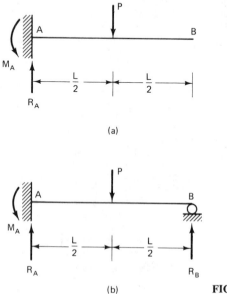

(a)

(b)   **FIGURE 11.1**

force-deflection relationships are written in such a manner that the forces are expressed in terms of the deformations, then the deformations become the unknowns in the analysis and the method is called a *deformation method*.

The method of consistent deformations to be presented in this chapter uses forces as unknowns and is accordingly designated a force method.

11.2 BASIC PRINCIPLES

To help us understand the basic principles involved in the method of consistent deformations, consider the beam in Fig. 11.2a. The first step in the analysis of this or any other indeterminate structure is to determine the degree of indeterminacy or the number of redundants that the structure possesses. As indicated in the figure, the beam has three unknown reactions. Since there are only two equations of equilibrium available for calculating the reactions, the beam is said to be indeterminate to the first degree. Looking at the situation from a different perspective, one can state that there exists one more reaction or restraint than is necessary to support the structure in a stable manner. For example, the cantilever beam that results if the vertical restraint R_B, at the right end of the beam, is removed suffices to support the load. Similarly, the simply supported beam obtained by removing the moment restraint M_A at the left end of the beam is also adequate for supporting the load. Restraints that can be removed without impairing the load-supporting capacity of the structure are referred to as *redundants*. In general the number of redundants that a structure possesses is equal to the degree of indeterminacy. The beam in Fig. 11.2a can thus be said to have one redundant or to be indeterminate to the first degree.

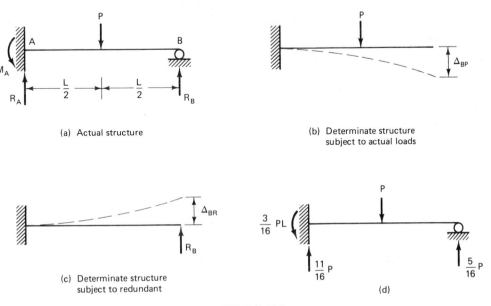

(a) Actual structure

(b) Determinate structure
subject to actual loads

(c) Determinate structure
subject to redundant

(d)

FIGURE 11.2

Having determined how many redundants a structure possesses, the next step is to decide which reaction is to be considered the redundant and to remove this restraint, thus forming a determinate structure. Any one of the reactions may be chosen to be the redundant provided that a stable structure remains after the removal of that reaction. For example, let us take the reaction R_B as the redundant. The determinate structure obtained by removing this restraint is the cantilever beam shown in Fig. 11.2b. We denote the deflection at end B of this beam, due to P, by Δ_{BP}. The first subscript indicates that the deflection is measured at B and the second subscript that the deflection is due to the applied load P. Using the moment-area method, it can be shown that $\Delta_{BP} = 5\,PL^3/48\,EI$.

Next we apply the redundant R_B to the determinate cantilever beam, as shown in Fig. 11.2c. This gives rise to a deflection Δ_{BR} at point B, whose magnitude can be shown to be $R_B L^3/3EI$.

In the actual indeterminate structure, which is subjected to the combined effects of the load P and the redundant R_B, the deflection at B is zero. Hence the algebraic sum of the deflection Δ_{BP} in Fig. 11.2b and the deflection Δ_{BR} in Fig. 11.2c must also vanish. Assuming downward deflections to be positive, we can write

$$\Delta_{BP} - \Delta_{BR} = 0 \tag{11.1}$$

or

$$\frac{5PL^3}{48EI} - \frac{R_B L^3}{3EI} = 0$$

from which

$$R_B = \frac{5}{16} P$$

Equation (11.1), which is used to solve for the redundant, is referred to as an equation of consistent deformations.

Once the redundant R_B has been evaluated, one can determine the remaining reactions by applying the equations of equilibrium to the structure in Fig. 11.2a. Thus $\sum F_y = 0$ leads to

$$R_A = P - \frac{5}{16}P = \frac{11}{16}P$$

and $\sum M_A = 0$ gives

$$M_A = \frac{PL}{2} - \frac{5}{16}PL = \frac{3}{16}PL$$

A free body of the beam, showing all the forces acting on it, is shown in Fig. 11.2d.

The method of consistent deformations, which has been presented here, can be summarized as follows. In the indeterminate structure (Fig. 11.2a) the deflection at point B is zero. When we remove the redundant R_B to obtain a determinate structure, we allow point B to deflect. We then ask: What value must the redundant R_B have so that its application to the determinate structure will cause point B to return to its original position of zero deflection? Viewed from a slightly different perspective, the method of consistent deformations can be considered to be an application of the principle of superposition. The structure in Fig. 11.2a can be assumed to be a determinate cantilever beam loaded with a known load P and an unknown load R_B. Since the deflection at B due to these two loads acting simultaneously is zero, the sum of the deflections that these loads produce at B when permitted to act one at a time must also vanish.

The individual steps in the method of consistent deformations can be summarized as follows:

1. Determine the number of redundants that the structure possesses.
2. Remove enough redundants to form a determinate structure. There is usually more than one way of doing this.
3. Calculate the displacements that the known loads cause in the determinate structure at the points where the redundants have been removed.
4. Calculate the displacements at these same points in the determinate structure due to the redundants.
5. At each point where a redundant has been removed, the sum of the displacements calculated in steps (3) and (4) must be equal to the displacement that exists at that point in the actual indeterminate structure. These relationships allow one to evaluate the redundants.
6. Knowing the values of the redundants, use equilibrium to detemine the remaining reactions.

11.3 APPLICATION OF CONSISTENT DEFORMATIONS TO STRUCTURES WITH ONE REDUNDANT

Example 11.1

Determine the reactions for the beam in Fig. 11.3a and draw its shear and moment diagrams.

As shown in the figure, the beam possesses three reactions, which is one more than needed to support the structure in a stable manner. Thus there exists one redundant reaction. Let us choose the reaction at B to be the redundant. Removal of R_B produces a determinate cantilever beam to which we apply separately the known load P (Fig. 11.3b) and the unknown redundant R_B (Fig. 11.3c). Since the actual beam in Fig. 11.3a, which is subjected to both P and R_B, has zero deflection at B, the sum of the deflections at B produced by P and R_B acting one at a time (Figs. 11.3b and 11.3c) must also vanish. Thus

$$\Delta_{BP} - \Delta_{BR} = 0 \qquad (11.2)$$

The deflections that appear in Eq. (11.2) can be determined using one of the methods presented in Chapters 6 and 7. In this instance we will use the moment-area method. Thus

$$\Delta_{BP} = \left(\frac{PL}{2EI}\right)(L)\left(\frac{L}{2}\right) + \left(\frac{PL}{EI}\right)\left(\frac{L}{2}\right)\left(\frac{2L}{3}\right) = \frac{7PL^3}{12EI}$$

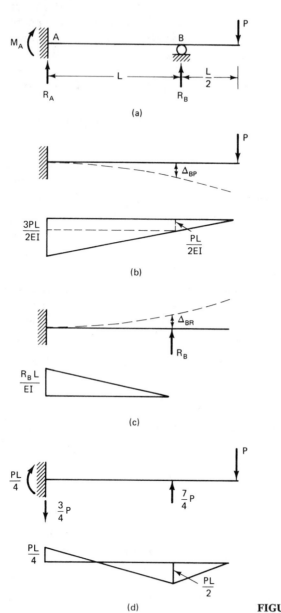

FIGURE 11.3

and

$$\Delta_{BR} = \left(\frac{R_B L}{EI}\right)\left(\frac{L}{2}\right)\left(\frac{2L}{3}\right) = \frac{R_B L^3}{3EI}$$

With these results Eq. (11.2) becomes

$$\frac{7PL^3}{12EI} - \frac{R_B L^3}{3EI} = 0$$

from which

$$R_B = \frac{7P}{4}$$

The remaining reactions are obtained by applying the equations of vertical and moment equilibrium to the free body in Fig. 11.3a. Thus

$$R_A = -\frac{3}{4}P, \qquad M_A = \frac{PL}{4}$$

A free body of the beam and its moment diagram are shown in Fig. 11.3d.

Example 11.2

When analyzing a structure by the method of consistent deformations one usually has a choice regarding the selection of the redundant. To illustrate this point, let us reanalyze the beam considered in Example 11.1, this time choosing M_A as the redundant instead of R_B. The beam is shown in Fig. 11.4a.

The determinate structure obtained by removing the moment restraint at A is a simply supported beam with an overhang on its right side. To this structure we

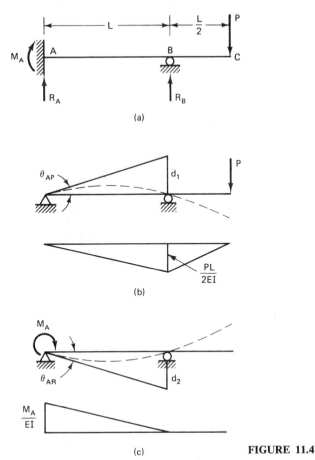

(a)

(b)

(c)

FIGURE 11.4

apply separately the applied load P, as shown in Fig. 11.4b, and the redundant M_A, as indicated in Fig. 11.4c. Since the slope at A is zero for the beam in Fig. 11.4a, the algebraic sum of the slopes at A, for the beams in Figs. 11.4b and 11.4c, must also vanish. Thus

$$-\theta_{AP} + \theta_{AR} = 0 \tag{11.3}$$

Using the moment-area method to determine the angles, we obtain for the beam in Fig. 11.4b

$$d_1 = \left(\frac{PL}{2EI}\right)\left(\frac{L}{2}\right)\left(\frac{L}{3}\right) = \frac{PL^3}{12EI}$$

$$\theta_{AP} = \frac{d_1}{L} = \frac{PL^2}{12EI}$$

and for the beam in Fig. 11.4c

$$d_2 = \left(\frac{M_A}{EI}\right)\left(\frac{L}{2}\right)\left(\frac{2L}{3}\right) = \frac{M_A L^2}{3EI}$$

$$\theta_{AR} = \frac{d_2}{L} = \frac{M_A L}{3EI}$$

Substituting the above results into Eq. (11.3), we obtain

$$-\frac{PL^2}{12EI} + \frac{M_A L}{3EI} = 0$$

and

$$M_A = \frac{PL}{4}$$

which is identical with the result obtained in Example 11.1.

Example 11.3

It is required to determine the reactions and to draw the bending-moment diagram for the beam in Fig. 11.5.

First we choose R_B as the redundant. Then we apply separately to the determinate beam, obtained by removing R_B, the 15 k load and the redundant R_B (Figs. 11.5b and 11.5c). Since the deflection at B is zero in the actual indeterminate structure,

$$\Delta_{BP} - \Delta_{BR} = 0 \tag{11.4}$$

The above deflections are determined using the conjugate beams shown in Figs. 11.5b and 11.5c. Thus

$$\Delta_{BP} = \frac{5833}{EI}$$

and

$$\Delta_{BR} = \frac{444.4 R_B}{EI}$$

Substitution of these values in Eq. (11.4) gives

$$R_B = 13.1 \text{ k}$$

A free body of the beam as well as its moment diagram are shown in Fig. 11.5d.

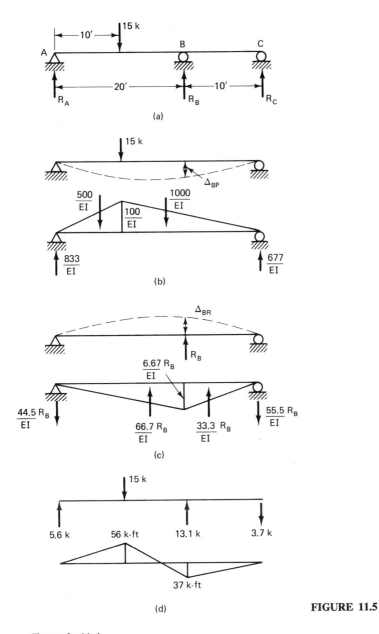

FIGURE 11.5

Example 11.4

It is required to determine the reactions and to draw the bending-moment diagram for the frame in Fig. 11.6a. Both members of the frame are assumed to have the same *EI*.

As shown in the figure, the frame has four reactions. The structure is thus indeterminate to the first degree and has one redundant reaction. Let us choose H_A

to be the redundant. To the determinate structure obtained by removing H_A we apply separately the uniformly distributed load, as indicated in Fig. 11.6b, and the redundant H_A, as shown in Fig. 11.6c. Since the horizontal deflection at A is zero in the actual structure, the sum of Δ_{AP} and Δ_{AR} must also vanish. Thus

$$\Delta_{AP} + \Delta_{AR} = 0 \tag{11.5}$$

The method of virtual work will be used to calculate both the magnitudes and the directions of the deflections. If we designate M_P as the moment due to the distributed load, M_R as the moment caused by the redundant H_A, and m as the moment due to a unit horizontal load at A (Fig. 11.6d), then

$$\Delta_{AP} = \int \frac{M_P m \, dx}{EI}$$

and

$$\Delta_{AR} = \int \frac{M_R m \, dx}{EI}$$

Substitution of the moments listed in the table in Fig. 11.6e into the relations above leads to

$$\Delta_{AP} = \int_0^{10} \frac{(25x - 2.5x^2)(-5 + 0.5x) \, dx}{EI}$$

$$\Delta_{AP} = \frac{-1042}{EI}$$

$$\Delta_{AR} = \int_0^5 \frac{H_A(x^2) \, dx}{EI} + \int_0^{10} \frac{(-5H_A + 0.5H_A x)(-5 + 0.5x) \, dx}{EI}$$

$$\Delta_{AR} = \frac{125H_A}{EI}$$

Hence Eq. (11.5) becomes

$$-1042 + 125H_A = 0$$

and

$$H_A = 8.34 \text{ kN}$$

A moment diagram of the frame is shown in Fig. 11.6f.

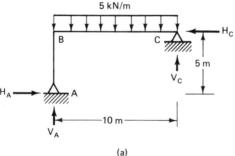

(a)   FIGURE 11.6

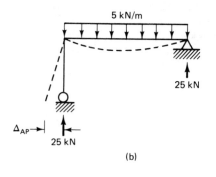

5 kN/m

25 kN

Δ_{AP}

25 kN

(b)

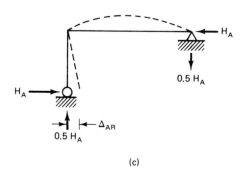

H_A

$0.5\,H_A$

H_A

$0.5\,H_A$

Δ_{AR}

(c)

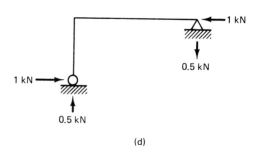

1 kN

0.5 kN

1 kN

0.5 kN

(d)

Element	x = 0	M_P	M_R	m
AB	A	0	$-H_A x$	$-x$
BC	B	$25x - 2.5x^2$	$-5H_A + 0.5H_A x$	$-5 + 0.5x$

Note: Moments are positive when they produce compression on the outside of the frame.

(e)

FIGURE 11.6 (Continued.)

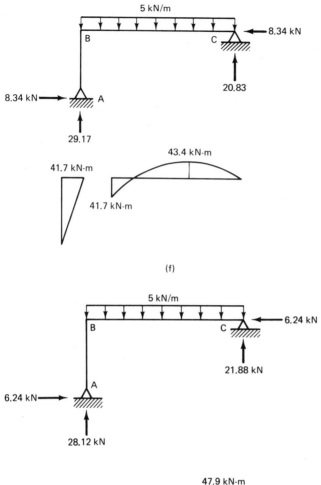

(f)

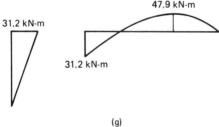

(g)   **FIGURE 11.6** (Continued.)

In the foregoing analysis it was assumed that both members of the frame had the same stiffness EI. Let us now repeat the analysis of the frame assuming this time that the stiffness of member BC remains equal to EI but that the stiffness of member AB is reduced to $0.5EI$.

Since Δ_{AP} depends only on the bending of member BC, whose stiffness is the same as it was before

$$\Delta_{AP} = \frac{-1042}{EI}$$

However, Δ_{AR} depends on the bending of both members AB and BC, and its value will now be given by

$$\Delta_{AR} = \int_0^5 \frac{H_A x^2 \, dx}{0.5EI} + \int_0^{10} \frac{(-5H_A + 0.5H_A x)(-5 + 0.5x)dx}{EI}$$

$$\Delta_{AR} = \frac{167H_A}{EI}$$

Substituting the new values of the deflections into Eq. (11.5) gives

$$-1042 + 167H_A = 0$$

from which

$$H_A = 6.24 \text{ kN}$$

The moment diagram corresponding to the new solution is given in Fig. 11.6g.

The above analysis and its results demonstrate an important characteristic of structural behavior: The internal load distribution in an indeterminate structure depends only on the relative values of EI of different parts of the structure, and not on the specific values of EI. In other words, the same reactions and moment diagrams would have been obtained regardless of the numerical values of E and I. The only facts that mattered were that EI had the same value for both members in the first part of the analysis and that EI of member BC was twice as large as the EI of member AB in the second part of the analysis.

Example 11.5

In each of the previous problems we considered elastic structures restrained by rigid supports. In other words, while the structure itself was able to deform, the supports remained stationary. Let us now analyze the beam in Fig. 11.7a, which is fixed at its left end and elastically restrained at point B by a cable.

The beam is indeterminate to the first degree, and we will let T, the force in the cable, be the redundant. Removal of the cable leads to a determinate cantilever beam, to which we apply separately the 5 kN force as shown in Fig. 11.7b and the redundant T as shown in Fig. 11.7c. In Example 11.1 we considered a beam similar to the one being analyzed here, except that the beam rested on an immovable support at B. As a consequence the sum of the deflections Δ_{BP} and Δ_{BR} was equal to zero. However, in the present case point B moves down by an amount equal to the elongation of the cable. Hence

$$\Delta_{BP} - \Delta_{BR} = \frac{TL}{AE} \tag{11.6}$$

Using the moment-area method, we obtain

$$\Delta_{BP} = \frac{630}{EI}, \qquad \Delta_{BR} = \frac{72T}{EI}$$

Substitution of these expressions together with values for A, E, and I into Eq. (11.6) gives

$$\frac{630 \times 10^9}{100 \times 10^6} - \frac{72T \times 10^9}{100 \times 10^6} = \frac{10T \times 10^3}{20} \tag{11.7}$$

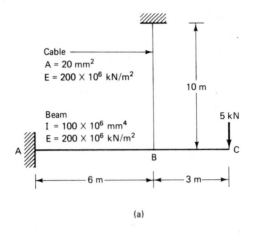

Cable
A = 20 mm²
E = 200 × 10⁶ kN/m²

Beam
I = 100 × 10⁶ mm⁴
E = 200 × 10⁶ kN/m²

10 m

5 kN

A

B

C

6 m

3 m

(a)

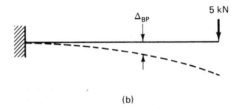

Δ_{BP}

5 kN

(b)

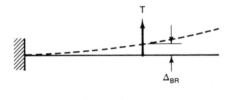

T

Δ_{BR}

(c)

FIGURE 11.7

from which

$$T = 5.16 \text{ kN}$$

If the support at B had been rigid, the right-hand side of Eq. (11.7) would have been zero and the reaction at B would have had the value

$$R_B = 8.75 \text{ kN}$$

Comparison of the values of R_B and T demonstrates that the rigid support takes more load than the elastic one.

11.4 SUPPORT SETTLEMENT

In general, the problem of support settlement is far more serious for indeterminate structures than it is for determinate ones. If one of the supports of a simply supported beam settles by a small amount, no major changes occur in the external or internal forces acting on the member. However, if one of the supports of a multispan beam settles a small amount, significant changes will occur in both the reactions and the bending moments. To see how the method of consistent deformations can be used to analyze a structure with support settlements, let us consider the following example.

Example 11.6

It is desired to determine the reactions and to draw the moment diagram for the beam in Fig. 11.8 assuming that the center support settles 2 in, and to compare the results thus obtained with those corresponding to zero settlement.

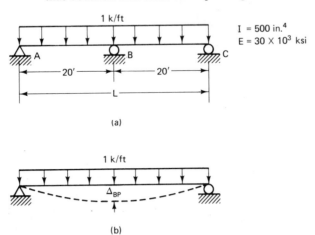

$I = 500$ in.^{4}
$E = 30 \times 10^3$ ksi

(a)

(b)

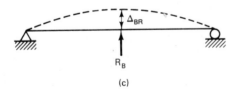

(c)

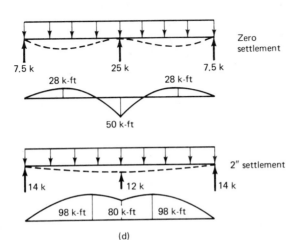

(d)

FIGURE 11.8

Let us choose the reaction at B as the redundant. To the determinate structure obtained by removing the redundant we apply first the distributed load as indicated in Fig. 11.8b and then reaction R_B as shown in Fig. 11.8c. In the absence of settlement at B,

$$\Delta_{BP} - \Delta_{BR} = 0 \tag{11.8}$$

However, if the support at B settles 2 in, then

$$\Delta_{BP} - \Delta_{BR} = 2 \tag{11.9}$$

Using the moment-area method, we obtain

$$\Delta_{BP} = \frac{5wL^4}{384EI} = \frac{(5)(1)(256)(10^4)(1728)}{(384)(30)(10^3)(500)} = 3.84 \text{ in}$$

and

$$\Delta_{BR} = \frac{RL^3}{48EI} = \frac{R(64)(10^3)(1728)}{(48)(30)(10^3)(500)} = 0.1536R \text{ in}$$

Substitution of these results into Eqs. (11.8) and (11.9) gives

$$R_B = 25 \text{ k} \quad \text{with zero settlement}$$

and

$$R_B = 12 \text{ k} \quad \text{with a 2 in settlement}$$

The moment diagrams corresponding to the two solutions are given in Fig. 11.8d. Comparison of the two diagrams indicates that the maximum moment increases from 50 k-ft to 98 k-ft as a result of the support settlement. It is thus evident that support settlement can give rise to significant increases in stress in an indeterminate structure.

11.5 STRUCTURES WITH SEVERAL REDUNDANTS

In previous sections of this chapter, the method of consistent deformations was used to analyze structures with one redundant. Let us now see how the same procedure can be applied to structures with two or more redundants. For example, the beam in Fig. 11.9a is indeterminate to the second degree and has two redundant reactions. If we let the reactions at B and C be the redundants, then the determinate structure obtained by removing these supports is the cantilever beam shown in Fig. 11.9b. To this determinate structure we apply separately the known distributed load (Fig. 11.9c) and the redundants R_1 and R_2 one at a time (Figs. 11.9d and 11.9e).

When a structure possesses several redundants, it is preferable to use numerical subscripts instead of letters for defining the redundants. Thus the reactions at B and C are referred to as R_1 and R_2. Furthermore, the deflections at these redundants will be denoted by Δ_1 and Δ_2. The association of deflections with redundants instead of points on the structure is a useful one because the deflections are always measured in the directions of the redundants.

Since the deflections at B and C in the original beam are zero, the algebraic sum of the deflections in Figs. 11.9c, 11.9d, and 11.9e at these same points must also vanish. Thus

$$\Delta_{1P} - \Delta_{11} - \Delta_{12} = 0 \tag{11.10}$$

$$\Delta_{2P} - \Delta_{21} - \Delta_{22} = 0 \tag{11.11}$$

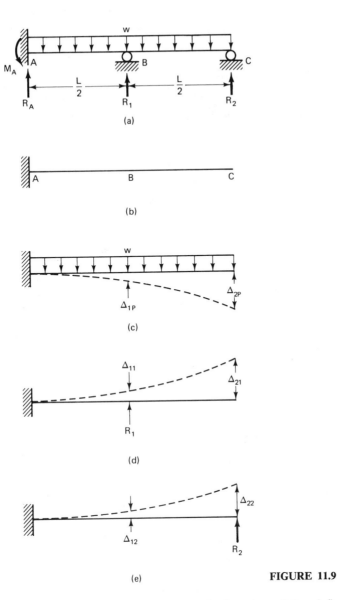

(a)

(b)

(c)

(d)

(e)

FIGURE 11.9

As before, the first subscript denotes the location of the deflection and the second refers to the force causing the deflection. For example, Δ_{12} is the deflection at R_1 due to R_2.

For complex structures it is useful to write the equations of consistent deformations in a form that differs slightly from the one used heretofore. Thus we let

$$\Delta_{ij} = \delta_{ij} R_j \tag{11.12}$$

where $\Delta_{ij} =$ the deflection at i due to redundant R_j
$\delta_{ij} =$ the deflection at i due to a unit load at j in the direction of R_j
$R_j =$ the redundant at j

Equation (11.12) states that the deflection at i due to a load at j is equal to the deflection at i due to a unit load at j multiplied by the value of the load at j.

Making use of the above notation, Eqs. (11.10) and (11.11) can be rewritten in the form

$$\Delta_{1P} - \delta_{11}R_1 - \delta_{12}R_2 = 0 \tag{11.13}$$

$$\Delta_{2P} - \delta_{21}R_1 - \delta_{22}R_2 = 0 \tag{11.14}$$

In this form, the equations allow one to distinguish clearly the known coefficients from the unknown variables.

Let us now consider some examples of structures with several redundants.

Example 11.7

It is required to find the reactions and to draw the moment diagram for the beam in Fig. 11.10a.

Since the structure is indeterminate to the second degree, we must remove two restraints to form a determinate structure. Let us choose the reactions at B and C as the two redundants and designate them as R_1 and R_2. We then apply to the determinate structure the known loads and each of the redundants, one at a time, as shown in Figs. 11.10b, 11.10c, and 11.10d. Since the deflections are zero at B and C in the original beam,

$$\begin{aligned} \Delta_{1P} - \delta_{11}R_1 - \delta_{12}R_2 = 0 \\ \Delta_{2P} - \delta_{21}R_1 - \delta_{22}R_2 = 0 \end{aligned} \tag{11.15}$$

Using the conjugate-beam method, we obtain

$$\Delta_{1P} = \frac{0.264PL^3}{EI}, \qquad \Delta_{2P} = \frac{0.215PL^3}{EI}$$

$$\delta_{11} = \delta_{22} = \frac{0.444L^3}{EI}, \qquad \delta_{12} = \delta_{21} = \frac{0.389L^3}{EI}$$

Substituting these values into Eqs. (11.15) and solving the resulting equations gives

$$R_1 = 0.725P, \qquad R_2 = -0.15P$$

A free body of the beam and its moment diagram are shown in Fig. 11.10e.

Example 11.8

It is required to determine the reactions and to draw the bending-moment diagram for the frame in Fig. 11.11a.

Let us choose the horizontal and vertical reactions at A as the redundants and label them R_1 and R_2 respectively. We then apply to the determinate structure, obtained by removing the redundants, the known distributed load, and the unknown redundants one at a time. The horizontal and vertical deflections at A due to these three force systems must vanish. Thus

$$\begin{aligned} \Delta_{1P} + \delta_{11}R_1 + \delta_{12}R_2 = 0 \\ \Delta_{2P} + \delta_{21}R_1 + \delta_{22}R_2 = 0 \end{aligned} \tag{11.16}$$

We will calculate the necessary deflections using the method of virtual work. Although for this structure it is possible to determine the direction of the deflections

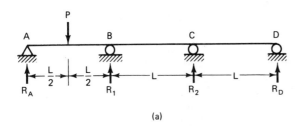

A P B C D

R_A $\frac{L}{2}$ $\frac{L}{2}$ R_1 L R_2 L R_D

(a)

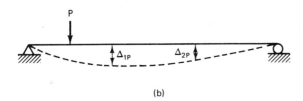

P

Δ_{1P} Δ_{2P}

(b)

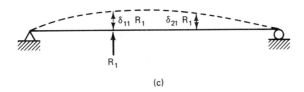

$\delta_{11}\,R_1$ $\delta_{21}\,R_1$

R_1

(c)

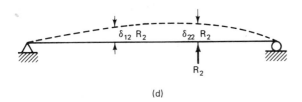

$\delta_{12}\,R_2$ $\delta_{22}\,R_2$

R_2

(d)

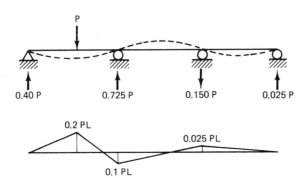

P

0.40 P 0.725 P 0.150 P 0.025 P

0.2 PL

0.025 PL

0.1 PL

(e)

FIGURE 11.10

by visual inspection, we will not do this. Instead, we will apply a more general procedure. We will initially assume all deflections to be in the same directions, namely,
the directions in which the redundants R_1 and R_2 have been assumed to act. Accordingly, we apply the unit loads to the left and upwards as indicated in Figs. 11.11c and
11.11d. Expressions for the bending moments due to the distributed load and due to
the unit loads are given in the table in Fig. 11.11e. They are designated as M_P, m_1,

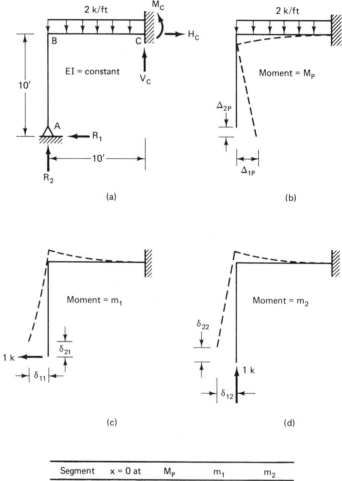

Segment	x = 0 at	M_P	m_1	m_2
AB	A	0	x	0
BC	B	$-x^2$	10	x

Note: Moments are positive when they
produce compression on the outside of
the frame

(e)

FIGURE 11.11

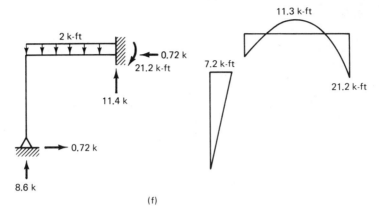

(f)

FIGURE 11.11 (Continued.)

and m_2, respectively. The moments are assumed to be positive when they produce compression on the outside of the frame.

Making use of the expressions in the table, we obtain

$$\Delta_{1P} = \frac{1}{EI} \int M_P m_1 \, dx = \int_0^{10} \frac{-10x^2 \, dx}{EI} = -\frac{3333}{EI}$$

$$\Delta_{2P} = \frac{1}{EI} \int M_P m_2 \, dx = \int_0^{10} \frac{-x^3 \, dx}{EI} = -\frac{2500}{EI}$$

$$\delta_{11} = \frac{1}{EI} \int m_1^2 \, dx = \int_0^{10} \frac{x^2 \, dx}{EI} + \int_0^{10} \frac{100 \, dx}{EI} = \frac{1333}{EI}$$

$$\delta_{12} = \delta_{21} = \frac{1}{EI} \int m_1 m_2 \, dx = \int_0^{10} \frac{10x \, dx}{EI} = \frac{500}{EI}$$

$$\delta_{22} = \frac{1}{EI} \int m_2^2 \, dx = \int_0^{10} \frac{x^2 \, dx}{EI} = \frac{333}{EI}$$

The negative signs of Δ_{1P} and Δ_{2P} indicate that the deflections are to the right and downward, respectively. These terms will therefore appear as negative quantities in Eqs. (11.16). By comparison, the remaining deflections are to the left and upward and appear as positive terms in the equations. Thus

$$-3333 + 1333R_1 + 500R_2 = 0$$
$$-2500 + 500R_1 + 333R_2 = 0$$

Solving these equations gives

$$R_1 = -0.72 \text{ k}, \qquad R_2 = 8.6 \text{ k}$$

The signs of these results indicate that R_1 acts to the right and R_2 upward.

Knowing R_1 and R_2, one can calculate the remaining reactions and plot a moment diagram for the frame as has been done in Fig. 11.11f.

11.6 STRUCTURES WITH INTERNAL REDUNDANTS

Each of the structures considered so far has been externally indeterminate. In other words, the structure possessed more reactions than there existed equations of equilibrium to solve for the reactions. However, it is also possible for a structure to be internally indeterminate. In such a case there are an insufficient number of equations of equilibrium to calculate all the internal forces in the structure, regardless of whether the external reactions can be determined or not.

Example 11.9

To see how an internally indeterminate structure is analyzed, let us consider the truss in Fig. 11.12a. It is assumed that EA is constant for all bars.

The truss has three reactions, which can readily be calculated using equations of equilibrium. However, if one then attempts to determine the bar forces, it soon becomes apparent that these cannot be obtained using only conditions of equilibrium. For example, if one applies the method of joints, it is evident that there exist at every joint three unknowns but only two equations of equilibrium. Since removal of any

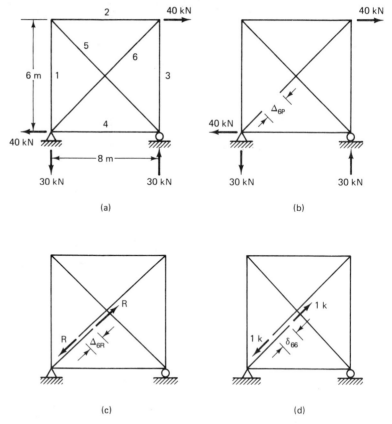

(a)

(b)

(c)

(d)

FIGURE 11.12

one of the bars changes the truss from an indeterminate structure to a determinate one, we conclude that the structure is indeterminate to the first degree.

Let us designate the force in bar 6 as the redundant and obtain a determinate structure by cutting this bar. To the determinate structure thus formed, we apply the known load of 40 kN as shown in Fig. 11.12b and the unknown redundant R, as indicated in Fig. 11.12c. Since in the actual structure there exists neither a gap nor an overlap in member 6, the sum of Δ_{6P} and Δ_{6R} must vanish. Thus

$$\Delta_{6P} + \Delta_{6R} = 0$$

or, in the alternate notation we have adopted,

$$\Delta_{6P} + \delta_{66}R = 0 \tag{11.17}$$

where δ_{66} is the deformation due to a unit force in bar 6 as shown in Fig. 11.12d.

To calculate the deformations appearing in the above equation we will use the method of virtual work. The data necessary to obtain the required deformations is

TABLE 11.1

Member	F (kN)	f (kN)	L (m)	FfL	f^2L
1	30	−0.6	6	−108	2.16
2	40	−0.8	8	−256	5.12
3	0	−0.6	6	0	2.16
4	40	−0.8	8	−256	5.12
5	−50	1.0	10	−500	10.0
6	0	1.0	10	0	10.0
				−1120	34.56

listed in Table 11.1. The symbols F and f represent the bar forces due to the applied load and the unit load in bar 6, respectively. Using the data in the table, we obtain

$$\Delta_{6P} = \Sigma \frac{FfL}{AE} = -\frac{1120}{AE}$$

$$\delta_{66} = \Sigma \frac{f^2L}{AE} = +\frac{34.56}{AE}$$

Substituting these values into Eq. (11.17) gives

$$-1120 + 34.56R = 0$$

from which

$$R = 32.41 \text{ kN}$$

The positive sign of R indicates that the direction assumed for the redundant, namely tension, was correct.

Once the force in bar 6 is known, the method of joints allows us to calculate the forces in the remaining bars. The results of these calculations are listed in Table 11.2.

TABLE 11.2

Member	Bar force (kN)
1	10.55 (T)
2	14.07 (T)
3	19.45 (C)
4	14.07 (T)
5	17.58 (C)
6	32.41 (T)

PROBLEMS

11.1 to 11.11. Solve for the reactions and draw the bending-moment diagram.

11.1.

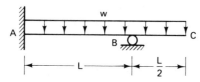

$I = $ constant

11.2.

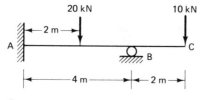

$I = $ constant

11.3.

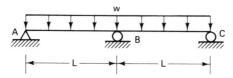

$I = $ constant

11.4.

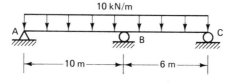

$I = $ constant

11.5.

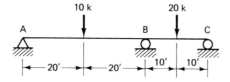

$I = $ constant

11.6.

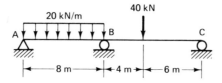

$I = $ constant

11.7.

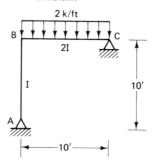

11.8.

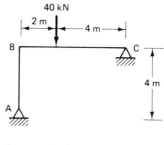

$I = $ constant

11.9. Analyze the frame in the figure assuming that **(a)** all members have the same I; **(b)** $I_{AB} = I_{CD} = 2I_{BC}$.

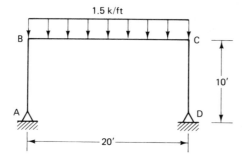

11.10.

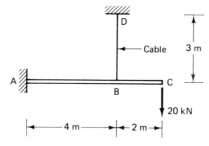

$I_{beam} = 100 \times 10^6$ mm⁴, $A_{cable} = 150$ mm², E $= 200 \times 10^6$ kN/m²

11.11.

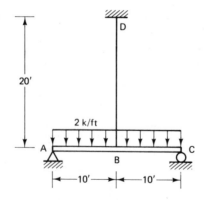

$I_{beam} = 72 \text{ in}^4 \qquad E = 30 \times 10^3 \text{ ksi}$
$A_{cable} = 0.1 \text{ in}^2$

11.12. Find the reactions assuming that the support at C settles 0.4 in.
$E = 30 \times 10^3$ ksi, $I = 1000 \text{ in}^4$

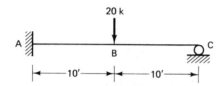

11.13. Assuming that the stiffness of the support at C is 4 k/in, find the reactions and determine the settlement at C.
$E = 30 \times 10^3$ ksi, $I = 1000 \text{ in}^4$

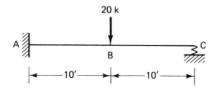

11.14 to 11.19. Solve for the reactions and draw the bending-moment diagram.

11.14.

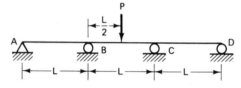

$I = \text{constant}$

11.15.

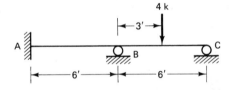

$I = \text{constant}$

11.16.

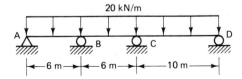

$I = \text{constant}$

11.17.

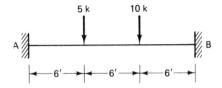

$I = \text{constant}$

11.18.

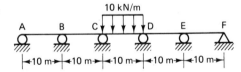

$I = \text{constant}$

11.19.

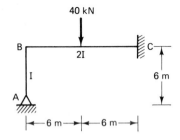

11.20. Determine all the bar forces.

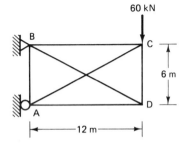

$A = \text{constant}$

11.21.   Determine all the bar forces.

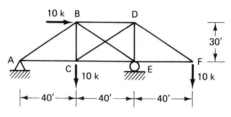

A = constant

12

Method
of Least Work

World Trade Center
Towers, N.Y.C., N.Y.
(*Courtesy of The Port
Authority of New York
and New Jersey.*)

12.1 INTRODUCTION

Castigliano's second theorem, sometimes called the method of least work, provides a useful alternative to the method of consistent deformations for analyzing certain types of indeterminate structures. Like the method of consistent deformations, the method of least work is a force method. In other words, the method uses forces called redundants as unknowns. The main difference between the two methods is that the strain energy of the system must be determined to formulate the equations in the method of least work, whereas deflections must be calculated to set up the governing equations in the method of consistent deformations. As a consequence the method of least work is preferable to the method of consistent deformations when it is easier to set up expressions for the strain energy of a system than it is to calculate displacements.

12.2 DERIVATION OF CASTIGLIANO'S THEOREMS

We will derive Castigliano's theorems in two stages. First we will formulate the law of reciprocal deformations, and then, using this principle, we will derive Castigliano's theorems.

To begin, let us consider the beam in Fig. 12.1a. Because of the load F_1 the beam deflects an amount Δ_1 at point 1 and an amount Δ_2 at point 2. Using the notation developed in Chapter 11, the deflections at 1 and 2 can be expressed in the form

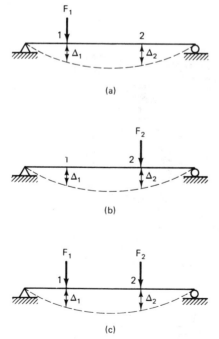

(a)

(b)

(c)   FIGURE 12.1

$$\Delta_1 = \delta_{11}F_1$$

$$\Delta_2 = \delta_{21}F_1$$

where δ_{11} and δ_{21} are the deflections at 1 and 2 due to a unit load at 1. If, instead of applying a load at 1, we apply a load at 2 as shown in Fig. 12.1b, the deflections at 1 and 2 can be written as

$$\Delta_1 = \delta_{12}F_2$$

$$\Delta_2 = \delta_{22}F_2$$

Using this notation, let us now formulate an expression for the work due to F_1 and F_2. It is assumed that F_1 and F_2 are applied one at a time and that the forces are applied gradually, so that both the forces and the deflections increase simultaneously from zero to their final values. If F_1 is applied first, the amount of work performed is

$$dW_1 = \tfrac{1}{2}F_1(\delta_{11}F_1)$$

Next we apply F_2 to the beam on which F_1 is already acting. The additional work resulting from the application of F_2 consists of two parts. First F_2 does an amount of work at point 2 equal to

$$dW_2 = \tfrac{1}{2}F_2(\delta_{22}F_2)$$

In addition, the force F_1, already on the beam, moves through the displacement that F_2 causes at point 1. Thus

$$dW_3 = F_1(\delta_{12}F_2)$$

The $\tfrac{1}{2}$ factor is absent in this expression because F_1 remains constant at its full value during the entire displacement.

The total work due to F_1 and F_2 is obtained by adding the increments of work due to the individual forces. Thus

$$W = \tfrac{1}{2}\delta_{11}F_1^2 + \tfrac{1}{2}\delta_{22}F_2^2 + \delta_{12}F_1F_2 \qquad (12.1)$$

Let us now determine the work if the loading process is reversed. That is, F_2 is applied first and F_1 is then applied to the beam on which F_2 is already acting. In this case the work due to F_2 is

$$dW_1 = \tfrac{1}{2}F_2(\delta_{22}F_2)$$

and the work due to F_1, applied next, is

$$dW_2 = \tfrac{1}{2}F_1(\delta_{11}F_1) + F_2(\delta_{21}F_1)$$

The total work due to F_1 and F_2 is now given by

$$W = \tfrac{1}{2}\delta_{11}F_1^2 + \tfrac{1}{2}\delta_{22}F_2^2 + \delta_{21}F_1F_2 \qquad (12.2)$$

In a linear system, the work performed by two forces is independent of the order in which the forces are applied. Hence the quantities given by Eqs. (12.1) and (12.2) must be equal, and it follows that

$$\delta_{12} = \delta_{21} \qquad (12.3)$$

The relationship given by Eq. (12.3) is known as *Maxwell's reciprocal theorem*.

Theorem 3. For a structure that behaves in a linear manner, the deflection produced at some point 1 by a unit load at some other point 2 is equal to the deflection at 2 due to a unit load at 1. The deflection at 1 is assumed to be in the same direction as the unit force at 1 and the deflection at 2 in the same direction as the unit force at 2.

Although it has not been demonstrated here, the theorem applies to moments and rotations as well as to forces and displacements. Thus the rotation at 1 due to a unit force at 2 would be equal to the deflection at 2 due to a unit couple at 1.

Having derived the law of reciprocal displacements, we are now in a position to derive Castigliano's theorems. With that end in mind let us redirect our attention to the beam we have been considering and this time apply the forces F_1 and F_2 simultaneously as shown in Fig. 12.1c. The resulting work is given by

$$W = \tfrac{1}{2}(F_1 \Delta_1 + F_2 \Delta_2) \tag{12.4}$$

where

$$\Delta_1 = \delta_{11}F_1 + \delta_{12}F_2$$
$$\Delta_2 = \delta_{21}F_1 + \delta_{22}F_2 \tag{12.5}$$

Substitution of the expressions in Eqs. (12.5) into Eq. (12.4) gives

$$W = \tfrac{1}{2}(\delta_{11}F_1^2 + \delta_{12}F_1F_2 + \delta_{21}F_1F_2 + \delta_{22}F_2^2)$$

Since the strain energy U stored in a deformed structure is equal to the work performed by the external loads, we can also write

$$U = \tfrac{1}{2}(\delta_{11}F_1^2 + \delta_{12}F_1F_2 + \delta_{21}F_1F_2 + \delta_{22}F_2^2)$$

In view of the law of reciprocal deformations given by Eq. (12.3), this expression for U can be simplified to

$$U = \tfrac{1}{2}(\delta_{11}F_1^2 + 2\delta_{12}F_1F_2 + \delta_{22}F_2^2)$$

Taking the derivative of U with respect to F_1, we obtain

$$\frac{\partial U}{\partial F_1} = \delta_{11}F_1 + \delta_{12}F_2$$

which in view of Eqs. (12.5) leads to

$$\frac{\partial U}{\partial F_1} = \Delta_1 \tag{12.6}$$

The relation given by Eq. (12.6) is known as *Castigliano's first theorem*.

Theorem 4. The partial derivative of the strain energy in a structure with respect to one of the external forces acting on the structure is equal to the displacement at that force in the direction of the force.

Equation (12.6) is often used to calculate deflections. More important, however, is the fact that it can also be used to analyze indeterminate structures. Thus if we consider the beam in Fig. 12.2 and apply Castigliano's theorem to find the deflection at R_1, which is of course zero, we obtain

$$\frac{\partial U}{\partial R_1} = 0 \tag{12.7}$$

Equation (12.7), is sometimes referred to as *Castigliano's second theorem*.

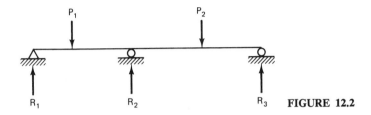

FIGURE 12.2

Theorem 5.   Redundants must have a value that will make the strain energy in the structure a minimum.

Accordingly, redundants in indeterminate structures can be determined by requiring that they minimize the strain energy in the system. The procedure of analyzing indeterminate structures in this way is called the method of least work.

12.3 APPLICATION OF THE METHOD OF LEAST WORK

To analyze an indeterminate structure, containing one or more redundants, by the method of least work, requires that one express the strain energy of the structure in terms of the redundants and then minimize the strain energy with respect to each of the redundants.

Strain energy is defined as the internal energy stored in a structure as a consequence of the deformation of the structure. It is obtained by calculating the work performed on the individual elements of the structure by the internal forces that act on the elements. Making use of Eqs. (7.4) and (6.21), we obtain for the strain energy of a structure consisting of flexural elements

$$U = \int \frac{M\,d\theta}{2} = \int \frac{M^2\,dx}{2EI} \tag{12.8}$$

and for the strain energy of a truss made up of axially loaded bars

$$U = \sum \frac{P\delta}{2} = \sum \frac{P^2 L}{2AE} \tag{12.9}$$

To see how the method of least work is applied to various structures, let us consider several examples.

Example 12.1

It is required to determine the reactions for the beam shown in Fig. 12.3a.

Since the beam has three reactions but only two equations of equilibrium are available to evaluate the reactions, the beam is indeterminate to the first degree and

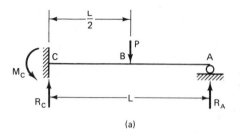

(a)

Element	x = 0	M	$\frac{\partial M}{\partial R_A}$
AB	A	$R_A x$	x
BC	A	$R_A x - P\left(x - \frac{L}{2}\right)$	x

(b)

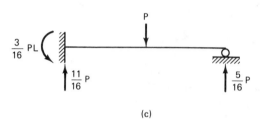

(c)

FIGURE 12.3

possesses one redundant reaction. Let us choose R_A as the redundant and determine its value by minimizing the strain energy in the beam with respect to R_A. That is,

$$\frac{\partial U}{\partial R_A} = 0 \qquad (12.10)$$

Since the strain energy in a beam is usually assumed to consist only of bending energy, it can be written as

$$U = \int_0^L \frac{M^2 \, dx}{2EI} \qquad (12.11)$$

To solve Eq. (12.10), one would ordinarily carry out the integration indicated by Eq. (12.11) first and then differentiate the resulting expression in accordance with Eq. (12.10). However, it is customary to reverse these procedures, that is, to differentiate first and then integrate. Thus

$$\frac{\partial U}{\partial R_A} = \int_0^L \frac{2M \dfrac{\partial M}{\partial R_A} \, dx}{2EI} = \int_0^L \frac{M \dfrac{\partial M}{\partial R_A} \, dx}{EI} = 0 \qquad (12.12)$$

It is evident from Eq. (12.12) that by differentiating first and then integrating one avoids squaring the moment in the strain energy term and thus reduces the numerical work significantly. The process of reversing the order of differentiation and integration used here can only be applied when the limits of integration are constant, as they are in our case.

Figure 12.3b lists expressions for the moment and its derivative needed in the solution of Eq. (12.12). It should be noted that the moment must be expressed as a function of the redundant R_A and that this is most easily accomplished by taking the origin of coordinates at A.

Substitution of the expressions in the table into Eq. (12.12) gives

$$\int_0^{L/2} \frac{R_A x^2 \, dx}{EI} + \int_{L/2}^{L} \left(R_A x^2 - P x^2 + \frac{PLx}{2} \right) dx = 0$$

from which one obtains

$$R_A = \frac{5}{16} P$$

A free body showing all the reactions appears in Fig. 12.3c.

Example 12.2

Determine the reactions for the beam shown in Fig. 12.4a.

The beam is indeterminate to the first degree, and we will designate R_A to be the redundant. The condition for determining R_A is

$$\frac{\partial U}{\partial R_A} = \int_0^L \frac{M \frac{\partial M}{\partial R_A}}{EI} \, dx = 0 \qquad (12.13)$$

To solve Eq. (12.13), the moment in the beam must be expressed as a function of R_A. Accordingly, it is necessary to express either R_C or R_D in terms of R_A. Since we will use only R_A and R_D in formulating the moment, we will determine R_D as a function of R_A. Taking moments about C, one obtains

$$20 R_A - 100 + 100 - 10 R_D = 0$$

$$R_D = 2 R_A$$

Figure 12.4b lists the expressions for the moment and its derivatives needed in the solution of Eq. (12.13). The origins used in the table were chosen so that the expressions for the moment are as simple as possible. Substitution of these expressions into Eq. (12.13) gives

$$\int_0^{10} R_A x^2 \, dx + \int_{10}^{20} (R_A x^2 - 10 x^2 + 100 x) \, dx + \int_0^{10} (4 R_A x^2 - 2 x^3) \, dx = 0$$

from which

$$R_A = 3.33 \text{ k}$$

A free body showing all the reactions appears in Fig. 12.4c.

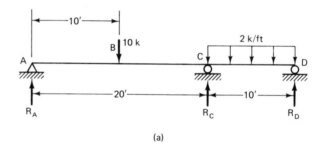

(a)

Element	x = 0	M	$\dfrac{\partial M}{\partial R_A}$
AB	A	$R_A x$	x
BC	A	$R_A x - 10(x - 10)$	x
CD	D	$2R_A x - x^2$	$2x$

(b)

10 k
2 k/ft

3.33 k
20 k
6.67 k

(c)

FIGURE 12.4

Example 12.3

It is required to determine the reactions for the frame shown in Fig. 12.5a. Assume that EI is constant.

Since the structure is indeterminate to the second degree, it has two redundant reactions. Let us choose R_1 and R_2, the reactions at A, to be the redundants. The conditions for evaluating R_1 and R_2 are

$$\frac{\partial U}{\partial R_1} = \int \frac{M \dfrac{\partial M}{\partial R_1} dx}{EI} = 0 \tag{12.14}$$

$$\frac{\partial U}{\partial R_2} = \int \frac{M \dfrac{\partial M}{\partial R_2} dx}{EI} = 0 \tag{12.15}$$

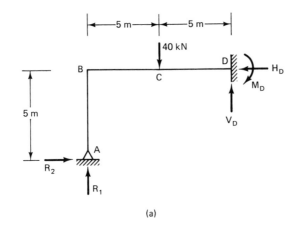

(a)

Element	x = 0	M	$\dfrac{\partial M}{\partial R_1}$	$\dfrac{\partial M}{\partial R_2}$
AB	A	$-R_2 x$	0	$-x$
BC	B	$R_1 x - 5R_2$	x	-5
CD	B	$R_1 x - 5R_2 - 40x + 200$	x	-5

(b)

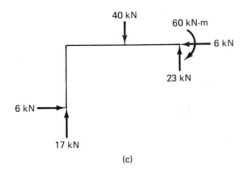

(c)

FIGURE 12.5

The expressions for the moment and its derivatives needed to solve Eqs. (12.14) and (12.15) are listed in Fig. 12.5b. Moments are considered positive if they cause compression on the outside of the frame.

Substitution of the expressions in Fig. 12.5b into Eqs. (12.14) and (12.15) gives

$$\int_0^5 (R_1 x^2 - 5R_2 x)\, dx + \int_5^{10} (R_1 x^2 - 5R_2 x - 40x^2 + 200x)\, dx = 0$$

$$\int_0^5 R_2 x^2\, dx + \int_0^5 (-5R_1 x + 25R_2)\, dx + \int_5^{10} (-5R_1 x + 25R_2 + 200x - 1000)\, dx = 0$$

227

from which

$$333R_1 - 250R_2 - 4167 = 0$$
$$-250R_1 + 292R_2 + 2500 = 0$$

and

$$R_1 = 17.0 \text{ kN}, \qquad R_2 = 6.0 \text{ kN}$$

A free body of the frame showing all the reactions appears in Fig. 12.5c.

Example 12.4

Determine the tension in the wires that support the beam shown in Fig. 12.6a. Each wire has a cross-sectional area of 0.1 in², $I = 288$ in⁴ for the beam, and $E = 30 \times 10^6$ psi for both the wires and the beam.

The structure is indeterminate to the first degree, and we will choose T_A, the force in the left-hand wire, to be the redundant. From equilibrium it is apparent that

$$T_C = T_A, \qquad T_B = 10 - 2T_A \tag{12.16}$$

The strain energy of the system consists of the energy of axial deformation in the wires and the bending energy in the beam. Thus

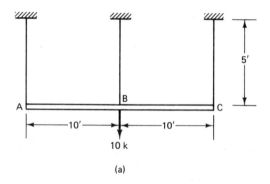

(a)

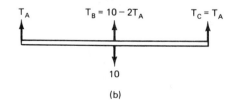

(b)

Element	x = 0	M	$\dfrac{\partial M}{\partial T_A}$
AB	A	$T_A x$	x
BC	C	$T_A x$	x

(c)

FIGURE 12.6

$$U = \int \frac{M^2 \, dx}{2EI} + \Sigma \frac{T^2 L}{2AE}$$

and

$$\frac{\partial U}{\partial T_A} = \int \frac{M \frac{\partial M}{\partial T_A}}{EI} \, dx + \Sigma \frac{T \frac{\partial T}{\partial T_A} L}{AE} = 0 \tag{12.17}$$

When we make use of the expressions in Eq. (12.16) and in Fig. 12.6c, Eq. (12.17) becomes

$$2 \int_0^{10} \frac{(T_A x)(x) \, dx}{EI} + \frac{2T_A(5)}{AE} + \frac{(10 - 2T_A)(-2)(5)}{AE} = 0$$

Substituting the values for A and I into the preceding expression and using the appropriate dimensional conversion factors, one obtains

$$\frac{667(1728)T_A}{288} + \frac{30(12)T_A}{0.1} = \frac{100(12)}{0.1}$$

from which

$$T_A = 1.58 \text{ k}$$

and

$$T_C = 1.58 \text{ k}, \qquad T_B = 6.84 \text{ k}$$

Example 12.5

Determine the bar forces for the truss shown in Fig. 12.7a.

Since the truss possesses one more bar than is necessary to resist the applied load in a stable manner, it is indeterminate to the first degree. Let us choose the force in bar AB as the redundant and designate it as R. Accordingly,

$$\frac{\partial U}{\partial R} = 0 \tag{12.18}$$

The strain energy for a structure consisting of axially loaded bars is

$$U = \Sigma \frac{S^2 L}{2AE} \tag{12.19}$$

where S is the force in any bar and the summation is carried out over all the bars. Substitution of Eq. (12.19) into Eq. (12.18) gives

$$\frac{\partial U}{\partial R} = \Sigma \frac{S \frac{\partial S}{\partial R} L}{AE} = 0 \tag{12.20}$$

Just as it was necessary to express the moment in a beam in terms of the redundant, we must now express the bar forces S in terms of the redundant R. Thus vertical equilibrium of joint A gives

$$S_{AD} = 1.34R - 89.4$$

and from horizontal equilibrium of joint A we obtain

$$S_{AC} = 80.0 - 2.0R$$

To evaluate R_1, using Eq. (12.20), the terms in the last column of Fig. 12.7b are added.

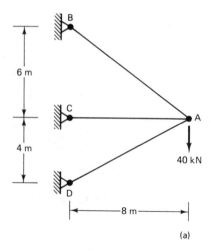

Member	A (cm²)	L (m)	S	$\dfrac{\partial S}{\partial R}$	$S\,\dfrac{\partial S}{\partial R}\,\dfrac{L}{A}$
AB	20	10	R	1	0.50 R
AC	10	8	80.0 − 2.0 R	−2.0	−128.0 + 3.20 R
AD	10	8.94	1.34 R − 89.4	1.34	−107.1 + 1.61 R
					−235.1 + 5.31 R

(b)

FIGURE 12.7

Thus

$$-235.1 + 5.31R = 0$$

from which

$$R = 44.3 \text{ kN}$$

and

$$S_{AC} = -8.6 \text{ kN}, \qquad S_{AD} = -30.0 \text{ kN}$$

Example 12.6

Draw the bending-moment diagram for the ring in Fig. 12.8a.

A closed ring is an indeterminate structure. This becomes evident if we consider the free body of half the ring, shown in Fig. 12.8b. From symmetry about a horizontal line through A and C we conclude that an axial force $P/2$ exists at both B and D and that the moments at B and D are equal. Furthermore, the ring is symmetric about a vertical line through B and D. As a consequence, shear, which is an asymmetric quantity, cannot be present at B and D. Since the use of symmetry is equivalent to employing equations of equilibrium, the above applications of symmetry have exhausted the available conditions of equilibrium and we are left with one unknown,

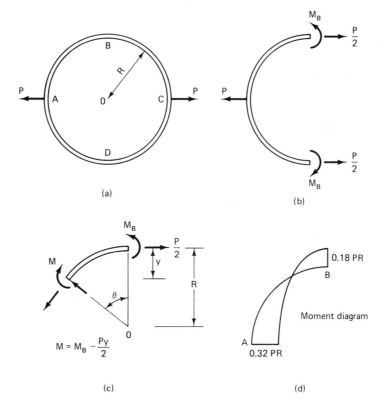

(a)

(b)

$$M = M_B - \frac{Py}{2}$$

(c)

0.18 PR

B

Moment diagram

A

0.32 PR

(d)

FIGURE 12.8

namely, M_B. The given ring is thus indeterminate to the first degree, and in accordance with Castigliano's theorem,

$$\frac{\partial U}{\partial M_B} = 0$$

Provided the depth of the ring cross section is small in comparison with the radius R of the ring, it is permissible to use the expression for the strain energy of a straight beam. Thus

$$U = \int \frac{M^2 \, ds}{2EI}$$

and

$$\frac{\partial U}{\partial M_B} = \int \frac{M \dfrac{\partial M}{\partial M_B} \, ds}{EI} = 0 \qquad (12.21)$$

It should be noted that the integration is carried out around the circumference of the ring.

When dealing with a circular member, polar coordinates are preferable to cartesian coordinates. If θ is measured counterclockwise from line OB, as shown in Fig. 12.8c, the moment at any section between B and A is given by

$$M = M_B - \frac{P}{2}y$$

$$= M_B - \frac{PR}{2}(1 - \cos \theta), \quad 0 < \theta < \frac{\pi}{2}$$

and

$$ds = R \, d\theta$$

Substitution of the above expressions into Eq. (12.21) and making use of symmetry gives

$$\frac{4}{EI} \int_0^{\pi/2} \left[M_B - \frac{PR}{2}(1 - \cos \theta) \right](1)R \, d\theta = 0$$

Carrying out the indicated integration leads to

$$M_B = \frac{PR}{\pi}\left(\frac{\pi}{2} - 1\right) = 0.18PR$$

from which

$$M = PR[0.18 - 0.5(1 - \cos \theta)], \quad 0 < \theta < \frac{\pi}{2}$$

A sketch of the moment diagram for one quadrant of the ring is shown in Fig. 12.8d.

PROBLEMS

12.1 to 12.12. Determine the reactions and draw the bending-moment diagram.

12.1.

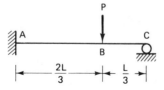

$I = $ constant

12.2.

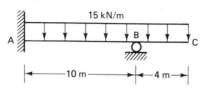

$I = $ constant

12.3.

$I = $ constant

12.4.

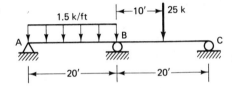

$I = \text{constant}$

12.5.

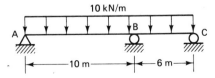

$I = \text{constant}$

12.6.

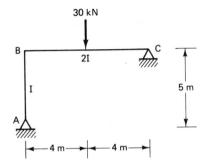

12.7.

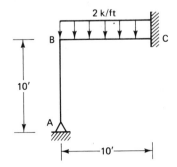

$I = \text{constant}$

12.8.

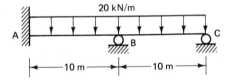

$I = \text{constant}$

12.9. $I_{beam} = 100$ in⁴, $A_{cable} = 0.1$ in².

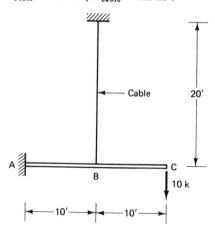

12.10. $I_{beam} = 180 \times 10^6$ mm⁴, $A_{cable} = 250$ mm².

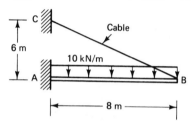

12.11. $I_{beam} = 576$ in⁴, $A_{cable} = 0.1$ in².

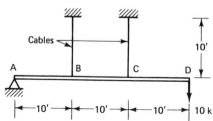

12.12.

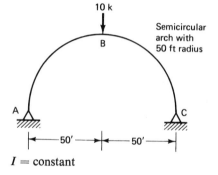

$I = $ constant

12.13. Draw the bending-moment diagram for the ring in the figure. I = constant

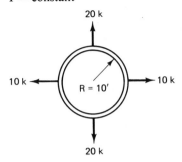

12.14. Draw the bending-moment diagram for the box frame in the figure. I = constant

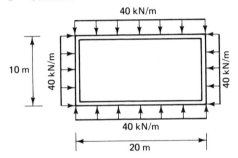

12.15. Find the bar forces for the truss in the figure. A = constant

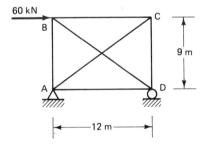

13

Slope-Deflection Method

13.1 INTRODUCTION

The methods of analysis considered so far, namely, the method of consistent deformations and the method of least work, use forces as unknowns. These methods are accordingly known as force methods. By comparison, the slope-deflection method to be considered in this chapter uses displacements as unknowns and is referred to as a displacement method. In the slope-deflection method, the moments at the ends of members are expressed in terms of the displacements and rotations of these ends.

An important characteristic of the slope-deflection method is that it does not become increasingly complicated to apply as the number of unknowns in the problem increases. In the slope-deflection method, as in the methods of consistent deformation and least work, an increase in the number of unknowns requires a corresponding increase in the number of equations that must be written and solved. However, whereas the complexity of each equation increases with the number of unknowns in the methods of consistent deformation and least work, in the slope-deflection method the individual equations are relatively easy to construct regardless of the number of unknowns. The slope-deflection method is thus fairly easy to apply even when the structure becomes relatively complex.

13.2 DERIVATION OF THE SLOPE-DEFLECTION EQUATION

When a rigid frame or a continuous beam is loaded, moments are developed at the ends of the individual members. The relationship that exists between these moments and the deformations at the ends of the members is the basis of the slope-deflection method. To derive this relationship, let us consider the member of constant EI shown in Fig. 13.1. It is assumed that the member in the figure is part of a rigid frame and that when loads are applied to the frame, the member will develop end moments and become deformed as indicated. The notation used in the figure to describe moments and deformations will be followed throughout the chapter.

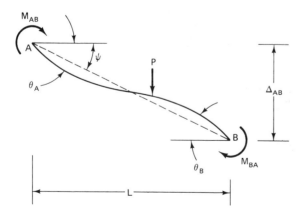

FIGURE 13.1

1. The moments at the ends of the member are designated as M_{AB} and M_{BA}, indicating that they act at ends A and B of member AB.
2. The rotations of ends A and B of the member are denoted by θ_A and θ_B. Since the rotations of all members of a rigid frame meeting at a common joint are equal, it is customary to refer to each of them as the joint rotation.
3. The term Δ_{AB} represents the translation of one end of the member relative to the other end, in a direction normal to the axis of the member. Sometimes the rotation of the axis of the member $\psi_{AB} = \Delta_{AB}/L$ is used in place of Δ_{AB}.

In dealing with the variables described above, the following sign convention is used.

1. The moments acting on the ends of the members are positive when clockwise.
2. The rotations of the ends of the members are also positive when clockwise.
3. The relative displacement of the ends of a member Δ is positive when the axis rotation ψ is clockwise.

Thus the moments and deformations are all positive when in the directions indicated in Fig. 13.1.

We will now derive an expression for the member moments M_{AB} and M_{BA} in terms of the deformations θ_A, θ_B, and Δ_{AB} and the load P acting on the member. It is easiest to carry out the proposed derivation by considering the effects of the four variables on the moments one at a time.

1. End moments due to rotation θ_A; $\theta_B = \Delta = P = 0$.

A member AB for which Δ and θ_B are zero can be represented by a beam simply supported at A and fixed at B as shown in Fig. 13.2. According to the second moment-area theorem, the distance between A and a tangent drawn to B is equal to the moment of the M/EI diagram between A and B about A. Thus

$$\frac{M_{AB}}{EI}\left(\frac{L}{2}\right)\left(\frac{L}{3}\right) - \frac{M_{BA}}{EI}\left(\frac{L}{2}\right)\left(\frac{2L}{3}\right) = 0$$

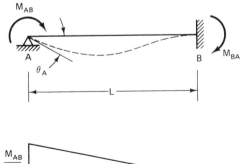

FIGURE 13.2

from which

$$M_{BA} = \frac{1}{2}M_{AB} \qquad (13.1)$$

From the first moment-area theorem, the change in slope between A and B is equal to the area of the M/EI diagram between A and B. Hence

$$\theta_A = \frac{M_{AB}}{EI}\left(\frac{L}{2}\right) - \frac{M_{BA}}{EI}\left(\frac{L}{2}\right)$$

In view of Eq. (13.1) this reduces to

$$M_{AB} = \frac{4EI\theta_A}{L}, \qquad M_{BA} = \frac{2EI\theta_A}{L} \qquad (13.2)$$

2. End moments due to rotation θ_B; $\theta_A = \Delta = P = 0$.

The member satisfying the above conditions is shown in Fig. 13.3. Since this beam is essentially identical to the one in Fig. 13.2,

$$M_{BA} = \frac{4EI\theta_B}{L}, \qquad M_{AB} = \frac{2EI\theta_B}{L} \qquad (13.3)$$

FIGURE 13.3

3. End moments due to a relative joint displacement Δ; $\theta_A = \theta_B = P = 0$.

Figure 13.4 depicts a member with a relative joint displacement but no joint rotations. From the first moment-area theorem, the change in slope between A and B

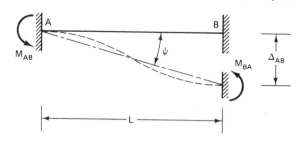

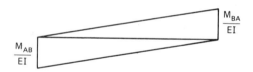

FIGURE 13.4

is given by the area of the M/EI diagram between A and B. Thus

$$\frac{M_{BA}}{EI}\left(\frac{L}{2}\right) - \frac{M_{AB}}{EI}\left(\frac{L}{2}\right) = 0$$

or

$$M_{BA} = M_{AB} \tag{13.4}$$

From the second moment-area theorem, the distance between B and a tangent drawn to A is equal to the moment of the M/EI diagram about B. Hence

$$\Delta = \frac{M_{BA}}{EI}\left(\frac{L}{2}\right)\left(\frac{L}{3}\right) - \frac{M_{AB}}{EI}\left(\frac{L}{2}\right)\left(\frac{2L}{3}\right)$$

Combining this expression with Eq. (13.4) gives

$$M_{AB} = M_{BA} = -\frac{6EI\Delta}{L^2} \tag{13.5}$$

The negative sign in Eq. (13.5) indicates that a positive translation leads to negative end moments. The translation Δ is positive because ψ is clockwise.

  4. End moments due to loads acting on the member; $\theta_A = \theta_B = \Delta = 0$.

The above conditions describe the fixed-end beam shown in Fig. 13.5. The moments M_{FAB} and M_{FBA} produced at the ends of such a member by loads acting along its span are accordingly called *fixed-end moments*. Table 13.1 gives the fixed-end moments for several common loading conditions. The fixed-end moments for other loads, not included in the table, can be derived using the moment-area method.

FIGURE 13.5

Combining Eqs. (13.2), (13.3), and (13.5) and the fixed-end moments, we obtain

$$M_{AB} = \frac{2EI}{L}\left(2\theta_A + \theta_B - \frac{3\Delta}{L}\right) \pm M_{FAB} \tag{13.6}$$

$$M_{BA} = \frac{2EI}{L}\left(2\theta_B + \theta_A - \frac{3\Delta}{L}\right) \pm M_{FBA} \tag{13.7}$$

  Equations (13.6) and (13.7) give the moments at ends A and B of a member AB in terms of the deformations at the ends of the member and in terms of any loads acting along the span of the member. These equations are known as the *slope-deflection equations*.

  If one compares Eqs. (13.6) and (13.7), it becomes apparent that the two equations are essentially identical. In other words, the moment at either end of a member is a function of twice the rotation at that end, the rotation at the far end, the relative joint displacement, and the fixed-end moment at the end being considered. Either one of the equations is thus sufficient to determine the moment at both ends of a member.

TABLE 13.1 FIXED-END MOMENTS

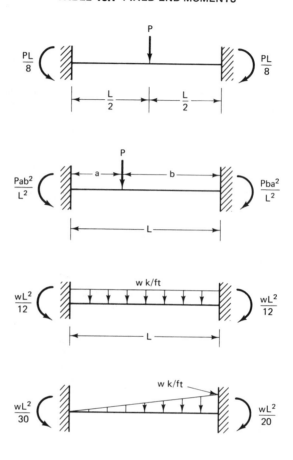

As was the case in the methods of consistent deformations and least work, when they were applied to flexural members, the slope-deflection equations consider only bending deformations. Deformations due to shear forces and axial loads in bending members are ignored. Since the deformations caused by shear forces and axial loads are very small compared to transverse bending deformations in most beams and frames, this is a reasonable assumption.

13.3 ALTERNATE DERIVATION OF SLOPE-DEFLECTION EQUATION

Instead of using the moment-area method to derive the slope-deflection equations, as was done in the preceding section, the slope-deflection equations can be derived using the differential equation of the member. Thus, let us consider the member in Fig. 13.6 and equate the internal moment EIy'' to the external moment at a distance

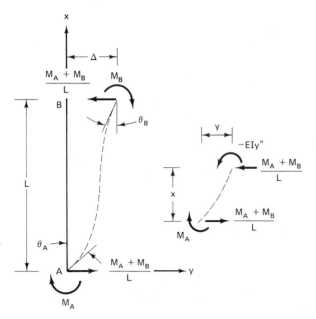

FIGURE 13.6

x from the lower end of the member. This leads to the differential equation

$$EIy'' = \frac{M_A}{L}(x - L) + \frac{M_B x}{L} \tag{13.8}$$

The solution of this equation is obtained by integrating both sides twice. Thus

$$EIy' = \frac{M_A}{L}\left(\frac{x^2}{2} - Lx\right) + \frac{M_B x^2}{2L} + C_1 \tag{13.9}$$

$$EIy = \frac{M_A}{L}\left(\frac{x^3}{6} - \frac{Lx^2}{2}\right) + \frac{M_B x^3}{6L} + C_1 x + C_2 \tag{13.10}$$

The boundary conditions at A are

$$y = 0 \quad \text{and} \quad y' = \theta_A \quad \text{at} \quad x = 0$$

Substitution of these conditions in Eqs. (13.10) and (13.9), respectively, gives

$$C_2 = 0 \quad \text{and} \quad C_1 = EI\theta_A$$

Next we apply the boundary conditions at B, which are

$$y' = \theta_B \quad \text{and} \quad y = \Delta \quad \text{at} \quad x = L$$

Substituting the first condition in Eq. (13.9) leads to

$$EI\theta_B = -\frac{M_A L}{2} + \frac{M_B L}{2} + EI\theta_A \tag{13.11}$$

and from the second condition and Eq. (13.10) we obtain

$$EI\Delta = -\frac{M_A L^2}{3} + \frac{M_B L^2}{6} + EI\theta_A L \tag{13.12}$$

Elimination of M_B between Eqs. (13.11) and (13.12) gives

$$M_A = \frac{2EI}{L}\left(2\theta_A + \theta_B - \frac{3\Delta}{L}\right) \qquad (13.13)$$

Similarly, we can eliminate M_A between Eqs. (13.11) and (13.12) and obtain

$$M_B = \frac{2EI}{L}\left(2\theta_B + \theta_A - \frac{3\Delta}{L}\right) \qquad (13.14)$$

Since no load was included along the span of the member in Fig. 13.6, Eqs. (13.13) and (13.14) give only the effect of joint deformations on the moments at the end of a member. The total end moments are obtained by adding the fixed-end moments, which are the effects of interior loads, to the above equations.

13.4 APPLICATION OF THE SLOPE-DEFLECTION METHOD

The primary aim of the slope-deflection method, like that of the methods of consistent deformations and least work, is to determine unknown reactions and internal forces. In the methods of consistent deformations and least work this is accomplished by setting up equations in terms of the unknown forces and then solving the equations for these forces. By comparison, the slope-deflection method employs a more indirect approach. Using the slope-deflection equation, the desired moments are first expressed in terms of unknown joint displacements. Equations of equilibrium are then written, and their solution gives the values of the joint displacements. Finally, the values of the joint displacements are substituted back into the slope-deflection equations, giving the desired moments. This indirect approach is justified because it leads to far less complex equations than those obtained using the more direct force methods.

Example 13.1

Determine the reactions and draw the moment diagram for the beam in Fig. 13.7a.

Unknowns. In the slope-deflection method, joint rotations and translations are the unknowns in terms of which the problem is formulated. None of the joints of the beam in Fig. 13.7a can translate, and only joint B can rotate. Hence θ_B is the only unknown, and we will need only one equation to determine its value.

Equations of Equilibrium. As a rule, unknown joint rotations are evaluated by writing equations of moment equilibrium at the joints that are free to rotate. Thus we write a moment-equilibrium equation for joint B, whose free body is depicted in

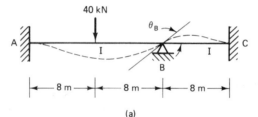

(a)   **FIGURE 13.7**

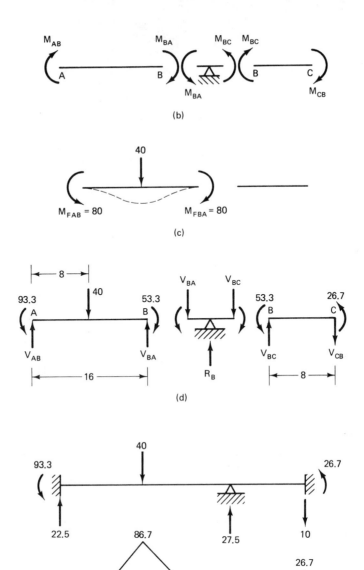

(b)

(c)

(d)

(e)

FIGURE 13.7 (Continued.)

Fig. 13.7b.

$$M_{BA} + M_{BC} = 0 \tag{13.15}$$

It is customary, in applying the slope-deflection method, to assume initially that all member moments are positive. In other words, the moments acting on the ends of the members are assumed to be clockwise as indicated in the figure. The directions of the moments acting on the joint then follow directly from this assumption.

Member Moments as Functions of Joint Displacements—Slope-Deflection Equations. Using the slope-deflection equation, the moments in Eq. (13.15) can be expressed in terms of the unknown θ_B. Since we will eventually wish to know the values of all the member moments, it is useful at this stage to write expressions not only for the two moments appearing in Eq. (13.15), but for all member moments. As a preliminary step, we calculate the fixed-end moments that are a part of the member moments.

$$M_{FAB} = -\frac{PL}{8} = -\frac{40(16)}{8} = -80$$

$$M_{FBA} = +80$$

As indicated in Fig. 13.7c, M_{FAB} is counterclockwise and M_{FBA} is clockwise. Since there are no loads on member BC, M_{FBC} and M_{FCB} are zero.

We are now ready to write expressions for all member moments using the slope-deflection equation. In view of the support conditions, the quantities, θ_A, θ_C, Δ_{AB}, and Δ_{BC} are all zero. Thus

$$M_{AB} = \frac{2EI}{16}(\theta_B) - 80 = 0.125EI\theta_B - 80$$

$$M_{BA} = \frac{2EI}{16}(2\theta_B) + 80 = 0.25EI\theta_B + 80$$

$$M_{BC} = \frac{2EI}{8}(2\theta_B) = 0.5EI\theta_B$$

$$M_{CB} = \frac{2EI}{8}(\theta_B) = 0.25EI\theta_B$$

Solution for Unknown Joint Displacements. Substitution of the above moments into Eq. (13.15) gives

$$0.75EI\theta_B + 80 = 0$$

from which

$$EI\theta_B = -106.7$$

Since use of the slope-deflection equation implies that we initially assume all θ's as well as all moments to be clockwise, the negative sign indicates that θ_B is counterclockwise.

Solution for Member Moments. Substitution of the solution for θ_B back into the expressions for the member moments gives

$$M_{AB} = 0.125(-106.7) - 80 = -93.3 \text{ kN-m}$$

$$M_{BA} = 0.25(-106.7) + 80 = +53.3 \text{ kN-m}$$

$$M_{BC} = 0.5(-106.7) = -53.3 \text{ kN-m}$$

$$M_{CB} = 0.25(-106.7) = -26.7 \text{ kN-m}$$

Reactions and Moment Diagrams. Having determined the member moments, we can now calculate the shears at the ends of the members and from these the reactions. This is accomplished using the free bodies shown in Fig. 13.7d. Note that M_{AB}, M_{BC}, and M_{CB}, which were found to be negative, are shown counterclockwise, while M_{BA}, which is positive, is shown clockwise. Taking moments about the left end of member AB gives

$$-V_{BA}(16) + 40(8) - 40.0 = 0$$

from which

$$V_{BA} = 17.5 \text{ kN}$$

Then from vertical equilibrium

$$V_{AB} = 22.5 \text{ kN}$$

In a similar manner one obtains

$$V_{BC} = V_{CB} = 10 \text{ kN}$$

The bending-moment diagram and the reactions are shown in Fig. 13.7e. The reaction at B is the algebraic sum of V_{BA} and V_{BC}.

The foregoing solution involves two separate and independent sign conventions, which should not be confused with one another. The first, employed in the slope-deflection equations, states that moments acting on the ends of members are positive if they are clockwise. This is a "rigid body" sign convention of the type used when Newton's law is used to write moment-equilibrium equations. By comparison, the second sign convention, used to construct the bending-moment diagram, states that moments producing compression on the upper side of a member are positive. This is a "strength of materials" sign convention that deals with stresses and deformations instead of external moments.

Example 13.2

Determine the reactions and draw the bending-moment diagram for the beam in Fig. 13.8a.

Unknowns. $\theta_A, \theta_B, \theta_C, \theta_D$.

Equations of Equilibrium.

$$\text{At joint A:}\quad M_{AB} = 0 \tag{13.16}$$

$$\text{At joint B:}\quad M_{BA} + M_{BC} = 0 \tag{13.17}$$

$$\text{At joint C:}\quad M_{CB} + M_{CD} = 0 \tag{13.18}$$

$$\text{At joint D:}\quad M_{DC} = 0 \tag{13.19}$$

Fixed-End Moments.

$$M_{FAB} = -\frac{20(12)^2(8)}{(20)^2} = -57.6$$

$$M_{FBA} = +\frac{20(8)^2(12)}{(20)^2} = +38.4$$

$$M_{FBC} = -\frac{2(20)^2}{12} = -66.7$$

$$M_{FCB} = +66.7$$

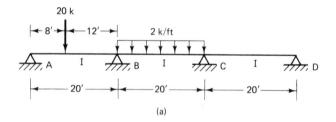

(a)

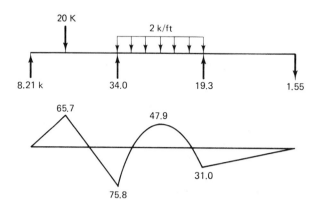

(b)   **FIGURE 13.8**

Member Moments.

$$M_{AB} = \frac{2EI}{20}(2\theta_A + \theta_B) - 57.6 = 0.2EI\theta_A + 0.1EI\theta_B - 57.6$$

$$M_{BA} = 0.2EI\theta_B + 0.1EI\theta_A + 38.4$$

$$M_{BC} = 0.2EI\theta_B + 0.1EI\theta_C - 66.7$$

$$M_{CB} = 0.2EI\theta_C + 0.1EI\theta_B + 66.7$$

$$M_{CD} = 0.2EI\theta_C + 0.1EI\theta_D$$

$$M_{DC} = 0.2EI\theta_D + 0.1EI\theta_C$$

At this stage one could simply substitute the above expressions for the member moments into the four equations of equilibrium and solve them for the unknown joint rotations. However, in a problem such as this, where one or more ends of the structure are simply supported, an alternative procedure, which simplifies the numerical work, can be employed. Before solving the equations of equilibrium for the unknown joint rotations, Eqs. (13.16) and (13.19) can be used to express θ_A and θ_D in terms of θ_B and θ_C, thereby eliminating two unknowns.

From Eq. (13.16) one obtains

$$EI\theta_A = 288 - 0.5EI\theta_B \tag{13.20}$$

and Eq. (13.19) gives

$$\theta_D = -0.5\theta_C \tag{13.21}$$

Solution for Unknowns. Substitution of the member moments together with relations (13.20) and (13.21) into Eqs. (13.17) and (13.18) gives

$$0.35EI\theta_B + 0.1EI\theta_C = -0.533$$
$$0.1EI\theta_B + 0.35EI\theta_C = -66.7$$

from which

$$EI\theta_B = 57.6, \qquad EI\theta_C = -206.9$$

Solutions for Member Moments.

$$M_{BA} = +75.8 \text{ k-ft}$$
$$M_{BC} = -75.8 \text{ k-ft}$$
$$M_{CB} = +31.0 \text{ k-ft}$$
$$M_{CD} = -31.0 \text{ k-ft}$$

The reactions and the bending-moment diagram corresponding to these member moments are shown in Fig. 13.8b.

Example 13.3

Determine the reactions and draw the bending-moment diagram for the structure in Fig. 13.9a.

Unknowns. θ_B, θ_C.

Equations of Equilibrium.

$$\text{At joint } B: \quad M_{BA} + M_{BC} + M_{BD} = 0 \tag{13.22}$$
$$\text{At joint } C: \quad M_{CB} - 27 = 0 \tag{13.23}$$

The joint at C has acting on it, in addition to the unknown member moment M_{CB}, a moment of -27 kN-m due to the load on the overhang CE.

Fixed-End Moments.

$$M_{FAB} = -\frac{12(10)^2}{12} = -100$$
$$M_{FBA} = +100$$
$$M_{FBC} = -\frac{6(10)^2}{12} = -50$$
$$M_{FCB} = +50$$

Member Moments.

$$M_{AB} = 0.4EI\theta_B - 100$$
$$M_{BA} = 0.8EI\theta_B + 100$$
$$M_{BC} = 0.4EI\theta_B + 0.2EI\theta_C - 50$$
$$M_{CB} = 0.4EI\theta_C + 0.2EI\theta_B + 50$$
$$M_{BD} = 0.667EI\theta_B$$
$$M_{DB} = 0.333EI\theta_B$$

Solution for Unknowns.   Using Eq. (13.23), we can express θ_C in terms of θ_B. Thus

$$EI\theta_C = -57.50 - 0.5EI\theta_B$$

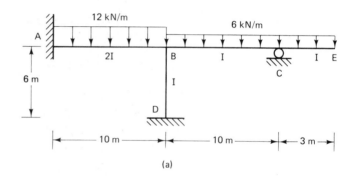

(a)

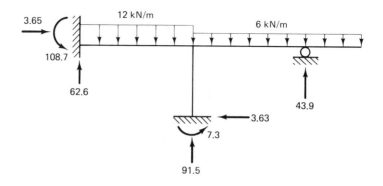

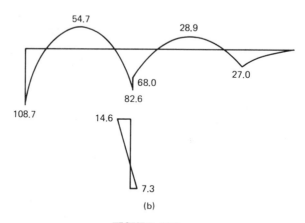

(b)

FIGURE 13.9

This result now makes it possible to evaluate θ_B in Eq. (13.22)

$$0.8EI\theta_B + 100 + 0.4EI\theta_B + 0.2(-57.50 - 0.5EI\theta_B) - 50 + 0.667EI\theta_B = 0$$

from which

$$EI\theta_B = -21.8, \qquad EI\theta_C = -46.6$$

Solutions for Member Moments.

$$M_{AB} = -108.7 \text{ kN-m}$$

$$M_{BA} = +82.6 \text{ kN-m}$$

$$M_{BC} = -68.0 \text{ kN-m}$$

$$M_{BD} = -14.6 \text{ kN-m}$$

$$M_{DB} = -7.3 \text{ kN-m}$$

The reactions and the moment diagram corresponding to these moments are shown in Fig. 13.9b.

Example 13.4

Determine the member moments and reactions and draw the bending-moment diagram for the frame in Fig. 13.10a.

Unknowns. $\theta_B, \theta_C, \Delta$.

A frame, whose upper part is not prevented from translating laterally, will sway either to the right or to the left a distance Δ, as shown in Fig. 13.10b. Since we ignore

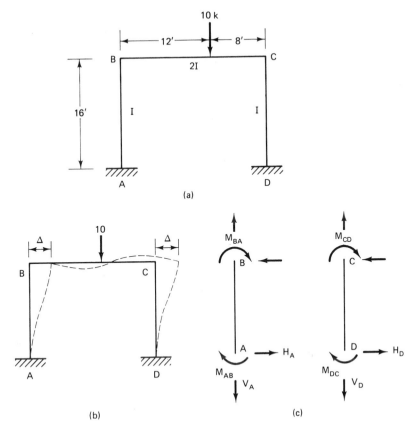

FIGURE 13.10

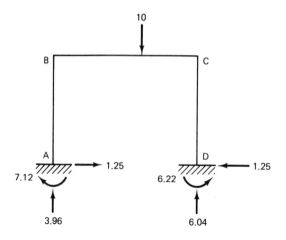

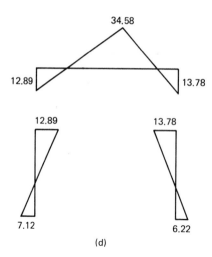

(d)

FIGURE 13.10 (Continued.)

axial shortening of flexural members, both that due to axial stress and that due to bending, ends B and C of member BC are assumed to move laterally the same amount, i.e., $\Delta_{BA} = \Delta_{CD} = \Delta$.

Equations of Equilibrium. The presence of the unknown Δ in addition to the unknown θ's requires that we employ an equation of horizontal equilibrium as well as equations of moment equilibrium. Thus

$$M_{BA} + M_{BC} = 0 \qquad (13.24)$$

$$M_{CB} + M_{CD} = 0 \qquad (13.25)$$

$$H_A + H_D = 0 \qquad (13.26)$$

Since the slope-deflection equation relates only member moments and not shears to the unknown joint rotations and displacements, it is necessary to express H_A and H_D in terms of the member moments. From Fig. 13.10c it is evident that

$$H_A = \frac{M_{AB} + M_{BA}}{16}, \qquad H_D = \frac{M_{DC} + M_{CD}}{16}$$

Consequently Eq. (13.26) can be rewritten as

$$M_{AB} + M_{BA} + M_{DC} + M_{CD} = 0 \tag{13.27}$$

Fixed-End Moments.

$$M_{FBC} = -\frac{10(64)(12)}{400} = -19.2$$

$$M_{FCB} = \frac{10(144)(8)}{400} = 28.8$$

Member Moments.

$$M_{AB} = 0.125EI\theta_B - 0.0234EI\Delta$$
$$M_{BA} = 0.250EI\theta_B - 0.0234EI\Delta$$
$$M_{BC} = 0.40EI\theta_B + 0.20EI\theta_C - 19.2$$
$$M_{CB} = 0.40EI\theta_C + 0.20EI\theta_B + 28.8$$
$$M_{CD} = 0.250EI\theta_C - 0.0234EI\Delta$$
$$M_{DC} = 0.125EI\theta_C - 0.0234EI\Delta$$

Solution for Unknowns.   Substitution of the foregoing member moments into the equations of equilibrium gives

$$0.650EI\theta_B + 0.20EI\theta_C - 0.0234EI\Delta = 19.2$$
$$0.20EI\theta_B + 0.650EI\theta_C - 0.0234EI\Delta = -28.8$$
$$0.375EI\theta_B + 0.375EI\theta_C - 0.0936EI\Delta = 0$$

from which we obtain

$$EI\theta_B = 46.1, \qquad EI\theta_C = -60.6, \qquad EI\Delta = -58.1$$

Solutions for Member Moments.

$$M_{AB} = 7.12 \text{ k-ft}$$
$$M_{BA} = 12.89 \text{ k-ft}$$
$$M_{BC} = -12.89 \text{ k-ft}$$
$$M_{CB} = 13.79 \text{ k-ft}$$
$$M_{CD} = -13.79 \text{ k-ft}$$
$$M_{DC} = -6.22 \text{ k-ft}$$

The reactions and the bending-moment diagram for the frame are shown in Fig. 13.10d.

PROBLEMS

13.1 to 13.12. Determine the member moments and the reactions and draw the bending-moment diagram.

13.1.

$I = $ constant

13.2.

$I = $ constant

13.3.

$I = $ constant

13.4.

13.5.

$I = $ constant

13.6.

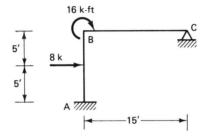

16 k-ft

B

5'

8 k

5'

A

15'

I = constant

13.7.

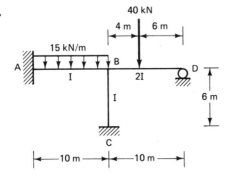

40 kN

4 m   6 m

15 kN/m

A   B   D

I 2I

I

6 m

C

10 m   10 m

13.8.

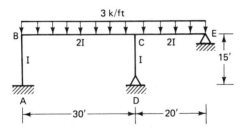

3 k/ft

B   E

2I C   2I

I I

15'

A   D

30'   20'

13.9.

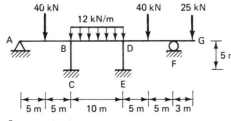

40 kN   40 kN   25 kN

12 kN/m

A   G

B   D

C   E   F

5 m

5 m 5 m   10 m   5 m 5 m 3 m

I = constant

13.10.

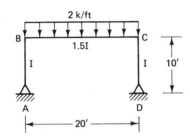

2 k/ft

B   C

1.5I

I I 10'

A   D

20'

255

13.11.

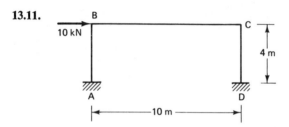

I = constant

13.12.

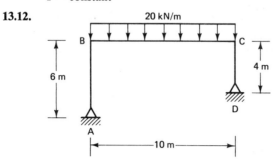

I = constant

14

Moment-Distribution Method

Newburg–Beacon Bridge over Hudson River, Newburg, N.Y. (*Courtesy of American Bridge Division, U.S. Steel Corporation.*)

14.1 INTRODUCTION

The moment-distribution method, like the slope-deflection method, is a deformation method. In other words, joint rotations and displacements are used as unknowns in carrying out the analysis. However, unlike the slope-deflection method or any of the other methods considered previously, the moment-distribution method does not require the solution of simultaneous equations. Instead, answers are obtained by a procedure of successive approximations, i.e., an iteration technique.

From Eqs. (13.6) and (13.7) it is evident that the moments acting on the ends of a member in a frame consist of several distinct parts. These include

1. The fixed-end moments, which are the moments caused by the loads acting on the member, with the ends of the member assumed to be fixed.
2. The moments due to the rotations that actually take place at the ends of the member.
3. The moments caused by the translation of one end of the member relative to the other.

In the moment-distribution method, this breakdown of member moments is utilized to arrive at a step-by-step procedure for calculating the value of the moments. First, all joints are assumed to be fixed and the external loads are applied; this gives rise to the so-called fixed-end moments. Next, those joints that can do so are permitted to rotate one at a time. This adds moments to the fixed-end moments in accordance with the joint rotations of the structure. Finally, the joints that are able to translate are allowed to do so, and a second correction is added to the existing member moments, bringing them to their final values.

The moment-distribution method and the slope-deflection method have the same theoretical basis. In both methods the moments at the ends of the members are considered to be functions of unknown joint rotations and translations. The methods differ only in the manner in which the solution is carried out. In the slope-deflection method all the unknowns are evaluated simultaneously by solving a set of equations, whereas in the moment-distribution method the solution is obtained by considering one unknown deformation at a time.

To see how the moment-distribution method is actually carried out, let us consider the two-span continuous beam in Fig. 14.1a. The beam, being fixed at A and C and free to rotate at B, will deform as shown by the dashed line and develop moments at A, B, and C as indicated. To determine the value of these moments, we initially assume the beam to be fixed at all joints as shown in Fig. 14.1b. This gives rise to fixed-end moments equal to 20 kip-ft at both ends of member AB and makes it necessary to apply an artificial external moment of 20 kip-ft to joint B. The external moment, called the *locking moment*, is necessary to prevent joint B from rotating. Since joint B is, of course, free to rotate in the actual structure, the next step in the procedure is to remove the external locking moment. This can be accomplished analytically by applying to joint B an external moment equal and opposite to the locking moment and then adding the effects of this unlocking moment to the already present member moments. The unlocking moment together with its effect on the individual members is depicted in Fig. 14.1c. It is evident from the free body of joint *B* that the unlocking moment is balanced by moments applied to the joint by members *AB* and *BC*. In other words, when joint *B* rotates, the rotation is resisted by the

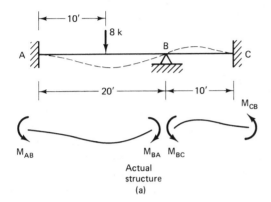

Actual
structure
(a)

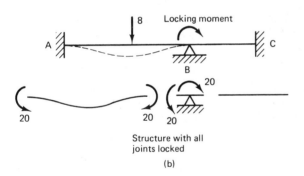

Structure with all
joints locked
(b)

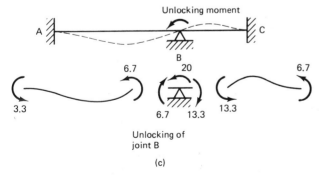

Unlocking of
joint B
(c)

$M_{AB} = 23.3$, $M_{BA} = 13.3$, $M_{BC} = 13.3$, $M_{CB} = 6.7$ **FIGURE 14.1**

members framing into the joint. The magnitudes of these resisting moments, which the individual members apply to the joint, are proportional to the stiffnesses of the members. Since the stiffness of a member is inversely proportional to its length, member *BC* applies a moment to joint *B* that is twice as large as the moment applied by member *AB*. Consideration of the members themselves indicates that the joint rotation at *B* induces member moments not only at *B* but also at *A* and *C*. This follows from the derivation presented on page 240, which demonstrated that rotation of one end of a member induces a moment at the far end one half as large as the moment at the near end, provided the far end is fixed. As a final step in the procedure, we add the member moments produced by the unlocking process to the fixed-end moments obtained while the joints were locked. This results in the member moments present in the actual structure.

In carrying out the solution of a problem by means of the moment-distribution method, it is unnecessary to consider both the member moments and the joint moments as was done here. Instead it suffices to calculate only the moments that act on the ends of the members. This includes the fixed-end moments that are caused by the applied loads while the joints are fixed and the additional moments induced by the joint rotations. Another important difference between the illustrative example we have just considered and the analysis of most actual structures is that the latter usually have not one but several joints that can rotate and possibly one or more joints that can translate. As a consequence, the analysis of a structure by the moment-distribution method usually involves considerably more numerical work than was required in the illustrative example.

14.2 BASIC CONCEPTS

To carry out the moment-distribution procedure requires the use of fixed-end moments, distribution factors, and carry-over moments. In addition, a sign convention must be adopted.

Sign convention.   Since the moment-distribution procedure is closely related to the slope-deflection method, it is desirable to use the same sign convention for the former as was used for the slope-deflection method in Chapter 13. Thus the moment acting on the end of a member is considered to be positive if it acts clockwise and negative if it acts counterclockwise.

Fixed-end moments.   As has already been explained in discussing the slope-deflection method, *fixed-end moments* are the moments that develop at the ends of a member as a result of the loads acting along the member if both ends of the member are fixed. Thus members having no loads acting along their span have zero fixed-end moments. The values for the fixed-end moments corresponding to some common loading conditions are given in Table 13.1.

Distribution factors.   In carrying out the moment-distribution method, joints are alternately locked and unlocked. When the joints are unlocked and permitted to rotate, the rotation is resisted by the members that frame into the joint. Thus the term *distribution factor* signifies that the resistance to the rotation of a joint is distributed among the members framing into the joint. Specifically, the distribution factor for any one member is a measure of the proportion of the total resistance to rotation supplied by that member.

To obtain an analytical expression for the distribution factor of a member, let us consider the joint shown in Fig. 14.2. The three members framing into the joint

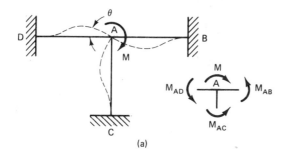

(a)

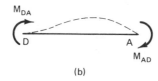

(b)   FIGURE 14.2

are rigidly connected to each other at the joint and fixed at their far ends. This condition always exists in the moment-distribution procedure, where all joints are fixed except the one that is being permitted to rotate. If a moment M is applied to joint A, the latter will rotate through an angle θ and resisting moments will develop because of the presence of the three members, as indicated in the figure. In accordance with the slope-deflection equation, Eq. (13.6), these moments are given by

$$M_{AB} = \frac{4EI_{AB}\theta}{L_{AB}}$$

$$M_{AC} = \frac{4EI_{AC}\theta}{L_{AC}}$$

$$M_{AD} = \frac{4EI_{AD}\theta}{L_{AD}}$$

It is customary to refer to the quantity I/L of a member as its *stiffness* and to denote this ratio by k. Thus the moments can be rewritten as

$$M_{AB} = 4E\theta k_{AB}$$
$$M_{AC} = 4E\theta k_{AC} \qquad (14.1)$$
$$M_{AD} = 4E\theta k_{AD}$$

From equilibrium of joint A,

$$M = M_{AB} + M_{AC} + M_{AD}$$

Substitution of (14.1) into this relation gives

$$M = 4E\theta \sum k_A \tag{14.2}$$

in which $\sum k_A$ represents the sum of the stiffnesses of the members framing into joint A.
If Eq. (14.2) is now rewritten in the form

$$4E\theta = \frac{M}{\sum k_A}$$

and this expression substituted back into Eqs. (14.1), we obtain

$$M_{AB} = \frac{Mk_{AB}}{\sum k_A}$$

$$M_{AC} = \frac{Mk_{AC}}{\sum k_A} \tag{14.3}$$

$$M_{AD} = \frac{Mk_{AD}}{\sum k_A}$$

It is evident from Eqs. (14.3) that an external moment applied to a joint will be resisted by the members framing into the joint in proportion to their stiffnesses k. The quantity $k/\sum k$, for a given member, is referred to as the *distribution factor* of that member. It gives the fraction of the total moment applied to a joint that is resisted by that member.

Carry-over moments.   If one considers the members framing into joint A, shown in Fig. 14.2a, it is evident that the rotation of joint A that takes place when the joint is unlocked produces member moments not only at A but also at B, C, and D. For example, member AD, depicted in Fig. 14.2b, develops a moment M_{DA} at D as well as a moment M_{AD} at A as a result of a rotation of joint A. The moment M_{DA} induced at the fixed end of member AD by the rotation at A is referred to as the *carry-over moment*, and the ratio of its value to M_{AD} is called the *carry-over factor*.

In Section 13.2 it was shown that, for a member such as AD, application of a moment M at the end that is free to rotate produces a moment at the fixed end having the same sign as M and half the magnitude of M. Thus

$$M_{DA} = \frac{1}{2}M_{AD}$$

The carry-over factor, which is defined as the ratio of the moment induced at the fixed end to the moment applied at the end that is able to rotate, is thus equal to $\frac{1}{2}$.

It should be kept in mind that the foregoing values of the carry-over factor, the distribution factors, and the fixed-end moments listed in Table 13.1 assume a constant stiffness for the entire length of the member.

14.3 APPLICATION OF MOMENT DISTRIBUTION TO STRUCTURES WITHOUT JOINT TRANSLATIONS

We will restrict our attention in the following examples to structures in which some or all of the joints are free to rotate but in which none of the joints are permitted to translate.

Example 14.1

It is required to determine the member moments and to plot the moment diagram for the beam in Fig. 14.3.

1. The first step is to calculate the distribution factors at all joints that are free to rotate. In this instance there is only one such joint, namely, joint B. The distribution factor of a member at a joint was shown in Section 14.2 to be equal to the stiffness k of the member divided by the sum of the k's of all the members framing into the joint. Thus

$$\text{D.F.}_{AB} = \frac{k_{AB}}{\sum k_B} = \frac{I/10}{3I/20} = \frac{2}{3}, \qquad \text{D.F.}_{BC} = \frac{I/20}{3I/20} = \frac{1}{3}$$

These factors are recorded in a table below the structure, as shown in the figure.

2. Next, assuming all joints to be locked, the fixed-end moments are calculated.

$$M_{FAB} = -\frac{PL}{8} = -\frac{20(10)}{8} = -25 \text{ kip-ft}, \qquad M_{FBA} = +25 \text{ kip-ft}$$

$$M_{FBC} = -\frac{wL^2}{12} = -\frac{1.5(400)}{12} = -50 \text{ kip-ft}, \qquad M_{FCB} = +50 \text{ kip-ft}$$

These quantities are also recorded in the table below the beam.

3. We are now ready to begin the actual moment-distribution process. Since joint B is the only joint that was artificially locked in the preceding step, it is the only joint that must now be unlocked. In the locked position an artificial external moment of 25 kip-ft was needed to keep the two member moments acting on joint B in equilibrium. When we release the joint, by removing the artificial locking moment, the moment necessary to produce equilibrium of the joint will be supplied by the members framing into the joint in proportion to their stiffnesses. In other words, the moment of $+25$ kip-ft needed to produce equilibrium is distributed among members AB and BC in accordance with the distribution factors. Thus member AB develops a moment of $+16.67$ kip-ft and member BC a moment of $+8.33$ kip-ft. These moments are recorded as shown in the figure, and a line is drawn below them to indicate that the joint is now in equilibrium.

4. According to the theory developed in Section 14.2, moments one half as large as those produced at end B of members AB and BC by the rotation of joint B are induced at ends A and C of these members. Thus moments of 8.33 kip-ft and 4.17 kip-ft are recorded at A and C in the table.

5. At this stage in the analysis joints A and C are fixed and joint B is externally unrestrained. Since this corresponds to the support conditions in the actual structure, the moment-distribution procedure is complete.

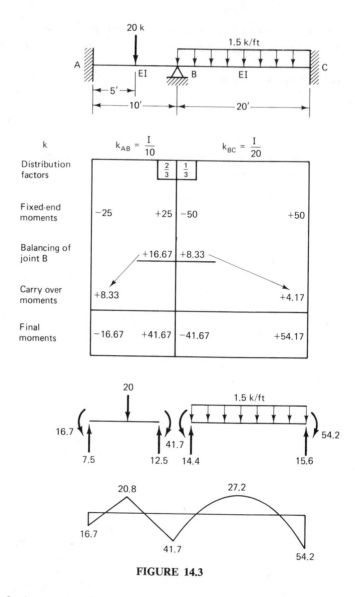

FIGURE 14.3

6. The final moment at the end of each member is obtained by adding the moments developed during each of the preceding steps. Thus

$$M_{AB} = -16.67 \text{ kip-ft}$$

$$M_{BA} = +41.67 \text{ kip-ft}$$

$$M_{BC} = -41.67 \text{ kip-ft}$$

$$M_{CB} = +54.17 \text{ kip-ft}$$

It should be evident from the foregoing analysis that whereas the moment-distribution process involves both members and joints, it is necessary in the actual calculations to record only the member moments.

Example 14.2

Determine the member moments and construct a moment diagram for the structure in Fig. 14.4.

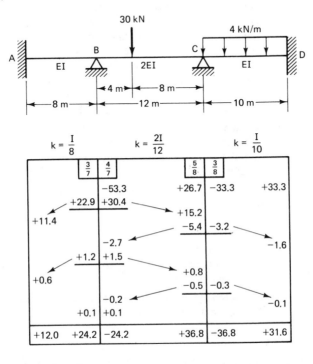

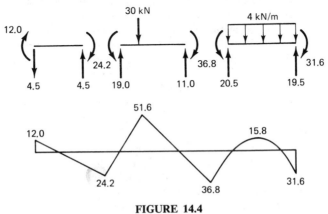

FIGURE 14.4

The main difference between this structure and the one considered in the previous example is that there are now two joints (B and C) that must be artificially locked at the outset and then released, whereas there was only one such joint in Example 14.1.

1. As before, the first step consists of calculating the distribution factors, locking all the joints, and determining the fixed-end moments. Since member AB is not subjected to any loads, there are no fixed-end moments at its ends.

2. We now proceed to unlock joints B and C one at a time. In other words, we unlock either one of these joints while holding the other fixed. Let us begin by unlocking joint B and distribute the resistance to the unbalanced moment of -53.3 kN-m to members AB and BC in accordance with the distribution factors. Thus member AB develops a moment of 22.9 kN-m and member BC a moment of 30.4 kN-m. Next we carry over one half of these moments to A and C and relock joint B. The line drawn under the moments at joint B signifies that the joint has been balanced and relocked.

3. We are now ready to proceed with the unlocking and balancing of joint C. The resistance to the unbalanced moment of 8.6 kN-m at C is distributed to members BC and CD, and one half of these moments is carried over to joints B and D. As before, the joint is relocked once these operations have been completed.

4. From here onward, we continue to carry out the process of unlocking, balancing, carrying over, and relocking first at one joint and then at the other. The procedure is stopped when the unbalanced moment remaining at both joints B and C is negligible.

   Unlike joints B and C, joints A and D are never released. As a consequence, no unbalanced moments are ever created at these joints and no carry over takes place.

5. The analysis is completed by adding the moments at the end of each member and thus obtaining the final member moments.

Example 14.3

Determine the member moments for the two-span beam in Fig. 14.5. The member is fixed at A, continuous over the support at B, and simply supported at C.

In calculating the distribution factor for member BC at joint C, it should be noted that BC is the only member framing into the joint. Hence, the distribution factor for member BC, which is equal to the ratio of k_{BC} to the sum of the k's of all the members framing into C, is equal to unity. This means that member BC must resist the entire unbalanced moment at C whenever the joint is released. By comparison, joint A is never released, and thus no unbalanced moment is ever created and no distribution factor is required. The distribution factors at B are determined in the usual manner.

We begin the analysis by locking joints B and C and calculating the fixed-end moments. Next joints B and C are alternately unlocked, balanced, and relocked. This procedure is continued until the unbalanced moment at both these joints becomes negligible. The final moment at the end of each member is the sum of the moments developed at that end during the preceding steps.

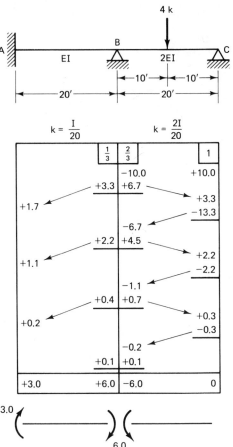

FIGURE 14.5

Simply Supported Ends

When there exists a simple support at the end of a structure, certain modifications can be introduced in the moment-distribution procedure to reduce the numerical work. In Section 14.2 it was shown that a moment applied to a joint at which several members meet, each of which is fixed at its far end, is distributed among these members in proportion to their stiffnesses. Let us now determine how this distribution of moments is affected if one of the members is hinged at its far end. Consider the three members meeting at joint A shown in Fig. 14.6. Members AB and AC are fixed at their far ends, while member AD is hinged at D. In accordance with Eq. (13.6), the member moments at A resulting from the application of an external moment M to joint A are

$$M_{AB} = 4Ek_{AB}\theta_A, \qquad M_{AC} = 4Ek_{AC}\theta_A, \qquad M_{AD} = 4Ek_{AD}\left(\theta_A + \frac{\theta_D}{2}\right) \qquad (14.4)$$

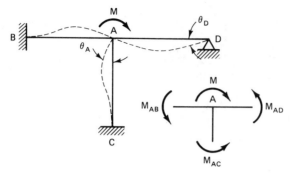

FIGURE 14.6

Since

$$M_{DA} = 4Ek_{AD}\left(\theta_D + \frac{\theta_A}{2}\right) = 0$$

it follows that

$$\theta_D = -\frac{\theta_A}{2}$$

and M_{AD} can be rewritten in the form

$$M_{AD} = 4E\theta_A\left(\frac{3}{4}k_{AD}\right) \tag{14.5}$$

Equilibrium of joint A requires that

$$M = M_{AB} + M_{AC} + M_{AD}$$

which, in view of Eqs. (14.4) and (14.5), can be written as

$$M = 4E\theta_A\left(k_{AB} + k_{AC} + \frac{3}{4}k_{AD}\right)$$

or simply

$$M = 4E\theta_A \sum k' \tag{14.6}$$

The term k', which will be referred to as the *effective stiffness* of a member, is equal to I/L if the member is fixed at its far end and $0.75I/L$ if the member is hinged at its far end.

From Eq. (14.6),

$$4E\theta_A = \frac{M}{\sum k'}$$

Substitution of this relation into Eqs. (14.4) and (14.5) gives

$$M_{AB} = \frac{Mk'_{AB}}{\sum k'}, \qquad M_{AC} = \frac{Mk'_{AC}}{\sum k'}, \qquad M_{AD} = \frac{Mk'_{AD}}{\sum k'} \tag{14.7}$$

Equations (14.7) indicate that the part of the total joint moment that is resisted by any one member framing into the joint is equal to $k'/\sum k'$. This result is identical to the one obtained in section 14.2 except that effective stiffnesses are now employed in place of ordinary stiffnesses.

It should also be noted that, whereas moments equal to $0.5M_{AB}$ and $0.5M_{AC}$ are induced at B and C by the application of M, no moment is induced at D. Thus the carry-over moment to a hinged end is zero.

To illustrate the concepts introduced above, we will make use of effective stiffnesses to reanalyze the beam considered in Example 14.3.

Example 14.4

Determine the member moments for the beam in Fig. 14.7 using effective stiffnesses. The effective stiffnesses for members AB and BC are

$$k'_{AB} = \frac{I}{20} = 0.05I, \qquad k'_{BC} = \frac{3}{4}\left(\frac{2I}{20}\right) = 0.075I$$

From these quantities the distribution factors shown in the figure are calculated. Next all joints are locked and the fixed-end moments determined.

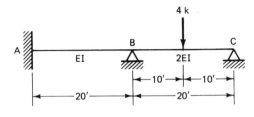

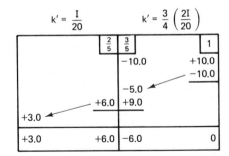

FIGURE 14.7

We are now ready to unlock the joints one at a time. Whenever effective stiffnesses are used, it is necessary to unlock all joints at simple supports first. Thus we begin by unlocking joint C, balancing the joint, and carrying over one half the balancing moment to B. If we were using ordinary stiffnesses, as we did in Example 14.3, we would now relock joint C and proceed to joint B. However, since the distribution factors at B are based on the assumption that joint C is hinged, we do not relock joint C prior to unlocking joint B. Furthermore, since joint C remains free to rotate during the remainder of the analysis, no moment is carried over to C when joint B is unlocked and balanced.

Example 14.5

Determine the member moments and draw the bending-moment diagram for the structure in Fig. 14.8.

The analysis of a frame like the one in Fig. 14.8, none of whose joints are per-

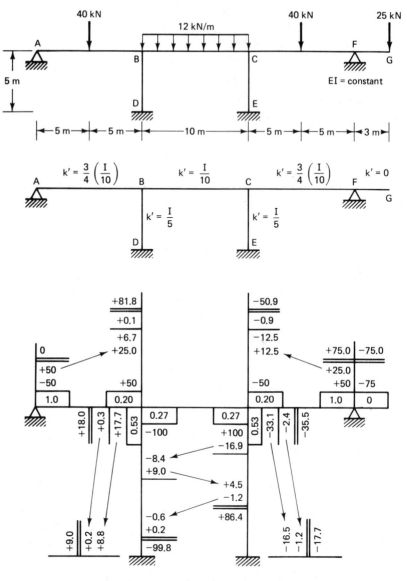

FIGURE 14.8

mitted to translate, is very similar to that of a continuous beam. However, it is necessary to record the moments in such a way that they do not interfere with one another. As will be demonstrated, this can be accomplished by recording some of the columns of moments vertically and some horizontally.

As usual, we begin by determining member stiffnesses, distribution factors, and fixed-end moments. Since simple supports exist at A and F, effective stiffnesses are used. The overhang to the right of F can offer no restraint to the rotation of joint F.

Hence both the effective stiffness of member *FG* and its distribution factor at *F* are zero. Furthermore the support at *F* is equivalent to a simple support as far as the effective stiffness of member *CF* is concerned.

    The use of effective stiffnesses requires that we begin the moment distribution process by unlocking and balancing joints *A* and *F*. These joints remain unlocked during the remainder of the procedure, which consists of alternately unlocking, balancing, and locking joints *B* and *C*.

14.4 STRUCTURES WITH JOINT TRANSLATIONS

In the preceding section moment distribution was applied solely to structures in which no joint translations occurred. We will now extend the method to structures having joints that can translate as well as rotate. For example, let us consider the frame in Fig. 14.9a. In this structure joints *B* and *C* rotate and translate laterally as indicated. When moment distribution is applied to a structure of this type, it is necessary to carry out the calculations in two stages. First a sufficient number of artificial restraints are added to the structure so that no joint translations can occur. For the structure being considered, this is accomplished by adding a horizontal restraint at *B* as shown in Fig. 14.9b. A routine moment distribution is then carried out for the artificially restrained structure. This leads to the member moments in the artificially restrained

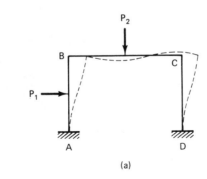

(a)

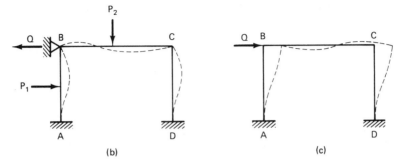

(b)   (c)

FIGURE 14.9

structure as well as the magnitude of the restraining force Q. A second set of calculations, whose pupose it is to cancel out the effects of the artificial restraint at B, is then performed. Accordingly, a force equal and opposite to Q is applied at B, as shown in Fig. 14.9c, and the member moments thus produced are added to those obtained in the previous stage of the analysis.

Example 14.6

Determine the member moments for the structure in Fig. 14.10a.

In the first stage of the analysis the structure is artificially restrained against lateral translation by the addition of a support at C. A routine moment distribution is then carried out for the artificially restrained structure. This analysis is presented in Fig. 14.10b.

The next step is to determine the magnitude of the artificial restraining force at C. Using the member moments determined in the preceding analysis and writing equations of equilibrium for the individual members of the frame (Fig. 14.10c), we obtain

$$F = -H_A + H_D = -\frac{14.2 + 28.4}{15} + \frac{22.7 + 45.5}{10} = 3.98 \text{ k}$$

Thus, the artificial restraint at C consists of a 3.98 k force acting to the right.

We are now ready to proceed with the second stage of the analysis, whose purpose is to cancel out the effects of the artificial restraining force F. Accordingly, we apply a force F', which is equal and opposite to F, to the structure as shown in Fig. 14.10d and determine the corresponding member moments. Since it is not possible to obtain a direct solution for member moments due to a given lateral force such as F', we proceed in an indirect manner as follows. If we assume all joints to be initially locked against rotation but free to translate, the lateral force F' gives rise to a lateral displacement Δ at B and C as indicated in Fig. 14.10d. The member moments that are produced when the ends of a member translate a distance Δ in relation to each other but do not rotate are, according to Eq. (13.5), given by

$$M_{AB} = M_{BA} = \frac{6EI\Delta}{(15)^2}$$

$$M_{CD} = M_{DC} = \frac{6EI\Delta}{(10)^2}$$

Not knowing the magnitude of Δ at this time, we arbitrarily assign values to the member moments that are proportional to the quantities I/L^2. Thus we let

$$M_{AB} = M_{BA} = 10 \text{ k-ft}$$

$$M_{CD} = M_{DC} = 22.5 \text{ k-ft}$$

These moments are fixed-end moments not unlike those used in the ordinary moment-distribution procedure. The difference is that the present moments are due to a joint translation whereas the previous ones were caused by loads acting normal to the members. However, in both instances they correspond to joints that have been prevented from rotating. As in the ordinary moment-distribution procedure, the next step therefore consists of releasing joints B and C, one at a time, balancing the joints, and recording the carry-over moments. During this procedure the translations of joints B

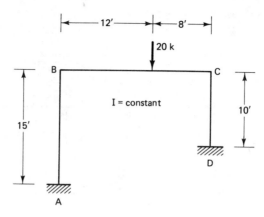

Actual structure

(a)

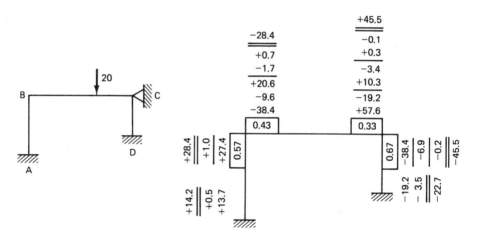

Analysis of artificially
restrained structure

(b)

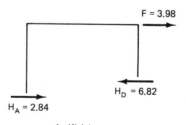

Artificial
restraining force

(c)

FIGURE 14.10

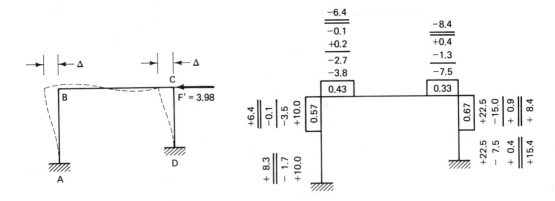

Structure subjected to sidesway

(d)

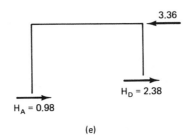

(e)

FIGURE 14.10 (Continued.)

and C are kept fixed at their original value, namely, Δ. The details of the moment distribution are recorded in Fig. 14.10d.

As before, we now determine the value of F' that corresponds to the member moments obtained in the analysis. Using the free body in Fig. 14.10e leads to

$$F' = H_A + H_D = \frac{8.3 + 6.4}{15} + \frac{15.4 + 8.4}{10} = 3.36 \text{ k}$$

It was our objective, in the second stage of the analysis, to determine moments corresponding to $F' = 3.98$ k. Instead, the moments we calculated correspond to a force of 3.36 k. However, member moments are proportional to the applied load, and the desired moments can therefore be obtained by multiplying the calculated ones by the ratio $3.98/3.36 = 1.18$.

The final member moments in the structure are now obtained by adding the corrected moments from the second analysis to the moments determined in the first analysis. Thus

$$M_{AB} = 14.2 + 1.18(8.3) = 24.0 \text{ k-ft}$$

$$M_{BA} = 28.4 + 1.18(6.4) = 36.0 \text{ k-ft}$$

$$M_{BC} = -28.4 + 1.18(-6.4) = -36.0 \text{ k-ft}$$

$$M_{CB} = 45.5 + 1.18(-8.4) = 35.6 \text{ k-ft}$$

$$M_{CD} = -45.5 + 1.18(8.4) = -35.6 \text{ k-ft}$$
$$M_{DC} = -22.7 + 1.18(15.4) = -4.5 \text{ k-ft}$$

PROBLEMS

14.1 to 14.10. Determine the member moments and the reactions and draw the bending-moment diagram.

14.1.

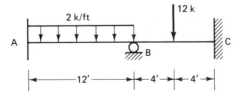

$I = \text{constant}$

14.2.

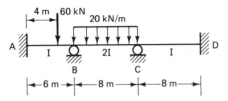

14.3.

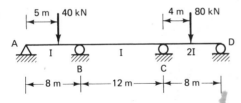

14.4.

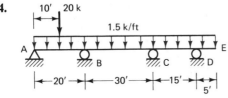

$I = \text{constant}$

14.5.

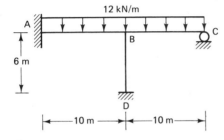

$I = \text{constant}$

14.6.

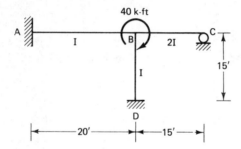

14.7.

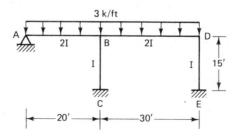

14.8.

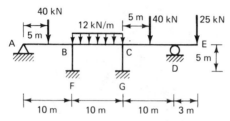

$I = \text{constant}$

14.9.

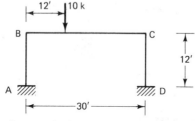

$I = \text{constant}$

14.10.

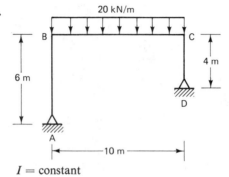

$I = \text{constant}$

15

Matrix Flexibility Method

Houston Astrodome, Houston, Tex. (*Courtesy of American Iron and Steel Institute.*)

15.1 INTRODUCTION

Matrix analysis is a systematic approach to the problem of analyzing large and complex structures with the help of an electronic computer. In matrix analysis, as in all other methods of indeterminate structural analysis, we form a set of simultaneous equations that describe the load-deformation characteristics of the structure being investigated. Matrix analysis differs from the other procedures in two basic ways. First, all calculations in matrix analysis are carried out with matrix algebra, and second, the aforementioned load-deformation characteristics of the structure are obtained from the load-deformation characteristics of discrete or finite elements into which the structure has been subdivided. Matrix algebra is ideally suited for setting up and solving equations on the computer, and the process of obtaining the behavior of a large structure from the behavior of a series of elements into which the structure has been subdivided greatly simplifies the analysis of even the most complex structures.

Matrix analysis is similar to the other methods we have considered in that the load-deformation characteristics can be formulated using either forces or displacements as the independent variables. The flexibility method, which we shall consider in this chapter, uses forces as the independent variables, and the stiffness method, to be considered in the next chapter, employs deformations as the independent variables. The two methods are consequently also referred to as the force method and the deformation method.

15.2 FLEXIBILITY MATRIX

The forces and the displacements that exist in a structure can be related to one another either by using flexibility or stiffness influence coefficients. The method of analysis to be considered in this chapter uses flexibility influence coefficients and is therefore referred to as the flexibility method.

Let us consider the beam shown in Fig. 15.1, along which we have chosen three points, designated as 1, 2, and 3. If we apply a load W_2 at point 2, the deflection Δ_1 at point 1 can be expressed in the form

$$\Delta_1 = F_{12}W_2$$

The term F_{12} is a *flexibility influence coefficient*. It is defined as the deflection at 1 due to a unit load at 2.

If, instead of a single load at 2, we apply loads at all three points, the deflection at 1 will be given by

$$\Delta_1 = F_{11}W_1 + F_{12}W_2 + F_{13}W_3$$

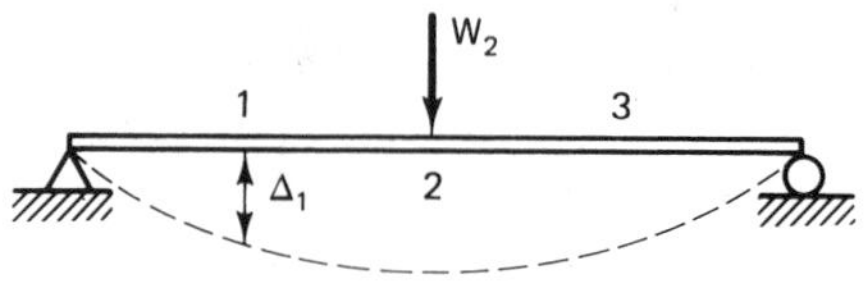

FIGURE 15.1

and the deflections at 2 and 3 by

$$\Delta_2 = F_{21}W_1 + F_{22}W_2 + F_{23}W_3$$
$$\Delta_3 = F_{31}W_1 + F_{32}W_2 + F_{33}W_3$$

These expressions can be rewritten in matrix form as

$$\begin{bmatrix} \Delta_1 \\ \Delta_2 \\ \Delta_3 \end{bmatrix} = \begin{bmatrix} F_{11} & F_{12} & F_{13} \\ F_{21} & F_{22} & F_{23} \\ F_{31} & F_{32} & F_{33} \end{bmatrix} \begin{bmatrix} W_1 \\ W_2 \\ W_3 \end{bmatrix}$$

or simply

$$[\Delta] = [F][W]$$

The matrix $[F]$ that contains the flexibility influence coefficients and relates the deflections $[\Delta]$ to the loads $[W]$ is referred to as the *flexibility matrix*. Maxwell's law of reciprocal deformations states that $F_{ij} = F_{ji}$. Thus the flexibility matrix is a symmetric matrix.

As an illustration, let us construct the flexibility matrix that relates the forces and displacements shown in Fig. 15.2a to one another. The vectors in the figure define not only the forces and deformations that we are considering but also their positive directions. Thus the vertical force W_1 and the corresponding vertical deflection Δ_1

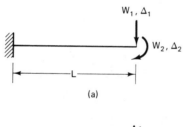

(a)

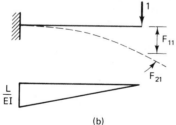

(b)

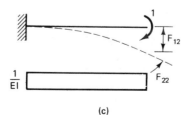

(c)

FIGURE 15.2

are positive when downward, and the moment W_2 and its corresponding rotation Δ_2 are positive when clockwise. It is customary to generalize the term "force" to include moments as well as forces and the term "deflection" to designate both rotations and translations.

The matrix equation that relates Δ_1 and Δ_2 to W_1 and W_2 is

$$\begin{bmatrix} \Delta_1 \\ \Delta_2 \end{bmatrix} = \begin{bmatrix} F_{11} & F_{12} \\ F_{21} & F_{22} \end{bmatrix} \begin{bmatrix} W_1 \\ W_2 \end{bmatrix} \tag{15.1}$$

According to the definition of a flexibility influence coefficient, F_{11} and F_{21} are the vertical deflection and rotation at the free end of the member due to a unit vertical load at that point. The first column of the flexbility matrix can thus be obtained by considering the beam in Fig. 15.2b. Using the moment-area method, we obtain

$$F_{21} = \frac{L^2}{2EI}, \qquad F_{11} = \frac{L^3}{3EI}$$

Similarly, the second column of the flexibility matrix is obtained by considering the beam in Fig. 15.2c. Thus

$$F_{12} = \frac{L^2}{2EI}, \qquad F_{22} = \frac{L}{EI}$$

Substituting the above influence coefficients into Eq. (15.1) gives

$$\begin{bmatrix} \Delta_1 \\ \Delta_2 \end{bmatrix} \begin{bmatrix} \dfrac{L^3}{3EI} & \dfrac{L^2}{2EI} \\ \dfrac{L^2}{2EI} & \dfrac{L}{EI} \end{bmatrix} \begin{bmatrix} W_1 \\ W_2 \end{bmatrix}$$

Example 15.1

Construct the flexibility matrix for the beam in Fig. 15.3a.

To obtain the first column in the matrix, we apply a unit vertical load at the free end of the member, as shown in Fig. 15.3b, and make use of the moment-area method. Thus

$$F_{11} = \frac{8L^3}{3EI}, \qquad F_{21} = \frac{2L^2}{EI}, \qquad F_{31} = \frac{5L^3}{6EI}, \qquad F_{41} = \frac{3L^2}{2EI}$$

In a similar manner columns 2, 3, and 4 are obtained by applying a unit moment at the free end and a unit force and a unit moment at midspan as shown in Figs. 15.3c, 15.3d, and 15.3e. The results, when combined, lead to the following flexibility matrix:

$$\begin{bmatrix} \Delta_1 \\ \Delta_2 \\ \Delta_3 \\ \Delta_4 \end{bmatrix} = \frac{1}{EI} \begin{bmatrix} \dfrac{8L^3}{3} & 2L^2 & \dfrac{5L^3}{6} & \dfrac{3L^2}{2} \\ 2L^2 & 2L & \dfrac{L^2}{2} & L \\ \dfrac{5L^3}{6} & \dfrac{L^2}{2} & \dfrac{L^3}{3} & \dfrac{L^2}{2} \\ \dfrac{3L^2}{2} & L & \dfrac{L^2}{2} & L \end{bmatrix} \begin{bmatrix} W_1 \\ W_2 \\ W_3 \\ W_4 \end{bmatrix} \tag{15.2}$$

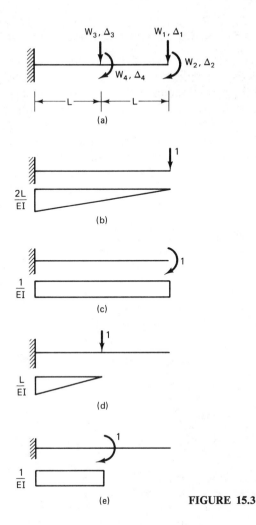

FIGURE 15.3

15.3 FORMATION OF THE STRUCTURE-FLEXIBILITY MATRIX FROM ELEMENT-FLEXIBILITY MATRICES

Although the method used to form the flexibility matrix in the previous section could theoretically be used on any structure, the difficulty of obtaining deflections in large and complex structures would make it extremely impractical. As a consequence, we will now introduce an alternative method for determining the flexibility matrix. In this method the structure is subdivided into several elements and a flexibility matrix is formed for each of the elements. The flexibility matrix for the entire structure is then obtained by combining the flexibility matrices of the individual elements. This process, whereby the behavior of a large system is synthesized from the behavior of the individual elements into which the system has been subdivided, is the basis of matrix analysis.

Element and Structure Forces and Deformations

Let us consider the beam in Fig. 15.4a. The four vectors in the figure define what we will refer to as the *structure forces and deformations*. They include two forces and two moments and the corresponding deflections and rotations. It is our aim to obtain the flexibility matrix that relates these quantities to each other, i.e., the matrix $[F]$ in the equation $[\Delta] = [F][W]$. To this end, we subdivide the beam into two elements, as indicated in Fig. 15.4b, and define two moments and their corresponding rotations for each element. A flexural element of a plane structure actually has six forces acting on it: an axial force, a shear, and a moment at each end. However, axial deformations are usually negligible in a flexural element, and only two of the remaining four forces are independent. Thus we have chosen the moments and the corresponding rotations at the ends of the elements as the forces and deformations we shall use to describe the behavior of the elements. To differentiate these quantities from the structure forces and deformations, we refer to them as *element forces and deformations*.

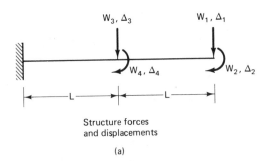

Structure forces
and displacements

(a)

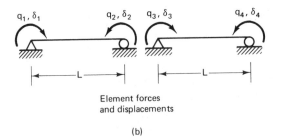

Element forces
and displacements

(b)

FIGURE 15.4

The juncture of two elements is called a *node*. For the time being we assume all structure loads to be applied only at nodes. In the flexibility method we may include as many or as few structure forces as we wish. The only restriction is that the forces be independent of one another. Thus reactions that are dependent on the applied loads cannot be included in the analysis. As far as element forces are concerned, they too must be independent. Thus we can use either the moments at both ends, as we have chosen to do, or the shear and the moment at one end.

Element-Flexibility Matrix

We shall now construct the flexibility matrix, which relates the deformations to the forces, for the typical flexural element shown in Fig. 15.5a. To obtain the first column of the matrix, we apply a unit moment $q_i = 1$ as shown in Fig. 15.5b. Using the moment-area method, we obtain

$$f_{ii} = \frac{d_1}{L} = \frac{L}{3EI}$$

and

$$f_{ji} = \frac{d_2}{L} = \frac{L}{6EI}$$

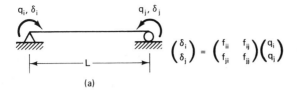

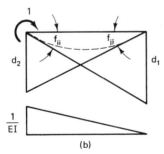

FIGURE 15.5

Similar results are obtained if we apply a unit moment to the right end of the element. The flexibility matrix thus takes the form

$$\begin{bmatrix} \delta_i \\ \delta_j \end{bmatrix} = \frac{L}{6EI} \begin{bmatrix} 2 & 1 \\ 1 & 2 \end{bmatrix} \begin{bmatrix} q_i \\ q_j \end{bmatrix} \tag{15.3}$$

Equation (15.3) gives the flexibility matrix for an element of a planar structure for which only flexural deformations are considered. It should be noted that lower-case letters are used when dealing with elements, and upper-case letters when the entire structure is being considered.

A great advantage of subdividing a structure into a series of elements is that the same element-flexibility matrix can be used for all the elements of the structure.

We shall now form a matrix containing a flexibility matrix of the type given by Eq. (15.3) for every element in the structure. Thus

$$\begin{bmatrix} \delta_1 \\ \delta_2 \\ \delta_3 \\ \delta_4 \end{bmatrix} = \frac{L}{6EI} \begin{bmatrix} 2 & 1 & 0 & 0 \\ 1 & 2 & 0 & 0 \\ 0 & 0 & 2 & 1 \\ 0 & 0 & 1 & 2 \end{bmatrix} \begin{bmatrix} q_1 \\ q_2 \\ q_3 \\ q_4 \end{bmatrix}$$

or

$$[\delta] = [f_c][q] \tag{15.4}$$

The matrix $[f_c]$ is called the *composite element-flexibility matrix*. It contains two element-flexibility matrices, one for each of the two elements into which the structure in Fig. 15.4 has been subdivided.

Force-Transformation Matrix

The composite flexibility matrix $[f_c]$ describes the force-deformation characteristics of the individual elements, taken one at a time. We shall now introduce a matrix that relates what occurs in these elements to the behavior of the entire structure. The matrix that performs this task in the flexibility method is known as the *force-transformation matrix*. Using the conditions of equilibrium, it relates the element forces to the structure forces. Thus

$$\begin{bmatrix} q_1 \\ q_2 \\ q_3 \\ q_4 \end{bmatrix} = \begin{bmatrix} B_{11} & B_{12} & B_{13} & B_{14} \\ B_{21} & B_{22} & B_{23} & B_{24} \\ B_{31} & B_{32} & B_{33} & B_{34} \\ B_{41} & B_{42} & B_{43} & B_{44} \end{bmatrix} \begin{bmatrix} W_1 \\ W_2 \\ W_3 \\ W_4 \end{bmatrix}$$

or

$$[q] = [B][W] \tag{15.5}$$

where $[B]$ is the force-transformation matrix.

As was done with the flexibility matrix, the force-transformation matrix will be formed one column at a time. Since the first column in $[B]$ contains the values of q_1 through q_4 due to a unit W_1, we obtain the terms in this column by applying a unit vertical load to the free end of the structure, as shown in Fig. 15.6a. Similarly, the remaining three columns of $[B]$ are obtained using the loadings shown in Figures 15.6b, 15.6c, and 15.6d. The element forces produced by each of the unit structure loads are shown in the figures. The negative signs indicate that positive structure loads cause negative element forces in the system we are considering.

The preceding results, when assembled, produce the following $[B]$ matrix:

$$\begin{bmatrix} q_1 \\ q_2 \\ q_3 \\ q_4 \end{bmatrix} = \begin{bmatrix} -2L & -1 & -L & -1 \\ -L & -1 & 0 & -1 \\ -L & -1 & 0 & 0 \\ 0 & -1 & 0 & 0 \end{bmatrix} \begin{bmatrix} W_1 \\ W_2 \\ W_3 \\ W_4 \end{bmatrix}$$

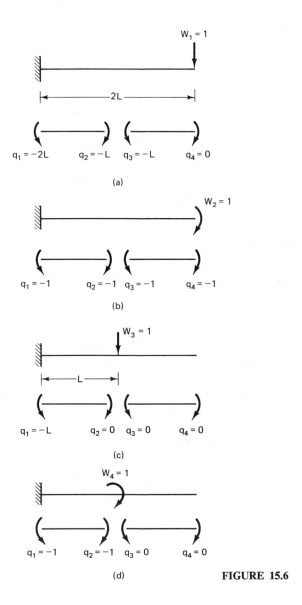

FIGURE 15.6

Transformation of $[f_c]$ into $[F]$

To develop the procedure for transforming the composite element-flexibility matrix $[f_c]$ into the structure-flexibility matrix $[F]$, we make use of the principle of conservation of energy. The work performed by the structure forces acting on the entire structure must be equal to the work of the element forces acting on all of the elements into which the structure has been subdivided. Thus

$$\frac{1}{2}[W_1 \quad W_2 \quad W_3 \quad W_4]\begin{bmatrix} \Delta_1 \\ \Delta_2 \\ \Delta_3 \\ \Delta_4 \end{bmatrix} = \frac{1}{2}[q_1 \quad q_2 \quad q_3 \quad q_4]\begin{bmatrix} \delta_1 \\ \delta_2 \\ \delta_3 \\ \delta_4 \end{bmatrix}$$

or

$$\frac{1}{2}[W]^{\mathrm{T}}[\Delta] = \frac{1}{2}[q]^{\mathrm{T}}[\delta]$$

Substitution of Eq. (15.4) for $[\delta]$ gives

$$[W]^{\mathrm{T}}[\Delta] = [q]^{\mathrm{T}}[f_c][q]$$

and, in view of Eq. (15.5), we obtain

$$[W]^{\mathrm{T}}[\Delta] = [W]^{\mathrm{T}}[B]^{\mathrm{T}}[f_c][B][W]$$

From which it follows that

$$[\Delta] = [B]^{\mathrm{T}}[f_c][B][W]$$

Comparison of this relation with

$$[\Delta] = [F][W]$$

indicates that

$$[F] = [B]^{\mathrm{T}}[f_c][B] \tag{15.6}$$

Equation (15.6) shows that the structure-flexibility matrix $[F]$ can be obtained from the composite element-flexibility matrix if the latter is premultiplied by $[B]^{\mathrm{T}}$ and postmultiplied by $[B]$. Since the synthesis of structure properties from those of the individual elements is the very essence of matrix analysis, one cannot overemphasize the importance of Eq. (15.6).

We will now use Eq. (15.6) to obtain the structure-flexibility matrix for the beam in Fig. 15.4.

$$[f_c][B] = \frac{L}{6EI}\begin{bmatrix} 2 & 1 & 0 & 0 \\ 1 & 2 & 0 & 0 \\ 0 & 0 & 2 & 1 \\ 0 & 0 & 1 & 2 \end{bmatrix}\begin{bmatrix} -2L & -1 & -L & -1 \\ -L & -1 & 0 & -1 \\ -L & -1 & 0 & 0 \\ 0 & -1 & 0 & 0 \end{bmatrix}$$

$$= \frac{L}{6EI}\begin{bmatrix} -5L & -3 & -2L & -3 \\ -4L & -3 & -L & -3 \\ -2L & -3 & 0 & 0 \\ -L & -3 & 0 & 0 \end{bmatrix}$$

$$[B]^{\mathrm{T}}[f_c][B] = \frac{L}{6EI}\begin{bmatrix} -2L & -L & -L & 0 \\ -1 & -1 & -1 & -1 \\ -L & 0 & 0 & 0 \\ -1 & -1 & 0 & 0 \end{bmatrix}\begin{bmatrix} -5L & -3 & -2L & -3 \\ -4L & -3 & -L & -3 \\ -2L & -3 & 0 & 0 \\ -L & -3 & 0 & 0 \end{bmatrix}$$

$$= \frac{1}{EI} \begin{bmatrix} \dfrac{8L^3}{3} & 2L^2 & \dfrac{5L^3}{6} & \dfrac{3L^2}{2} \\[2ex] 2L^2 & 2L & \dfrac{L^2}{2} & L \\[2ex] \dfrac{5L^3}{6} & \dfrac{L^2}{2} & \dfrac{L^3}{3} & \dfrac{L^2}{2} \\[2ex] \dfrac{3L^2}{2} & L & \dfrac{L^2}{2} & L \end{bmatrix}$$

The preceding flexibility matrix is seen to be identical to that obtained in Example 15.1.

Example 15.2

Construct the structure-flexibility matrix for the frame in Fig. 15.7a.

Both structure and element forces are defined in the figure. On the basis of these, the composite element-flexibility matrix $[f_c]$ and the force-transformation matrix $[B]$ can be formed. Thus

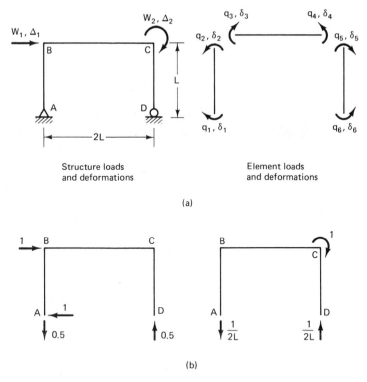

Structure loads
and deformations

Element loads
and deformations

(a)

(b)

FIGURE 15.7

$$[f_c] = \frac{L}{6EI}\begin{bmatrix} 2 & 1 & & & & \\ 1 & 2 & & & & \\ & & 4 & 2 & & \\ & & 2 & 4 & & \\ & & & & 2 & 1 \\ & & & & 1 & 2 \end{bmatrix}$$

and

$$\begin{bmatrix} q_1 \\ q_2 \\ q_3 \\ q_4 \\ q_5 \\ q_6 \end{bmatrix} = \begin{bmatrix} 0 & 0 \\ L & 0 \\ L & 0 \\ 0 & -1 \\ 0 & 0 \\ 0 & 0 \end{bmatrix}\begin{bmatrix} W_1 \\ W_2 \end{bmatrix}$$

The free bodies used to construct $[B]$ are shown in Fig. 15.7b.

The structure-flexibility matrix is formed using the relation $[F] = [B]^T[f_c][B]$. Hence

$$\begin{bmatrix} \Delta_1 \\ \Delta_2 \end{bmatrix} = \frac{L}{6EI}\begin{bmatrix} 6L^2 & -2L \\ -2L & 4 \end{bmatrix}\begin{bmatrix} W_1 \\ W_2 \end{bmatrix}$$

where $[F]$ is the 2×2 matrix on the right side of the equation.

Example 15.3

Construct the structure-flexibility matrix for the truss in Fig. 15.8a.

The structure forces are defined in Fig. 15.8a, and the element forces are assumed to consist of a single tension force for each bar. The element-flexibility matrix for an axially loaded bar of the type shown in Fig. 15.8b is given by

$$[\delta_i] = \left[\frac{L}{AE}\right][q_i]$$

Accordingly, the composite element-flexibility matrix takes the form

$$[f_c] = \frac{1}{AE}\begin{bmatrix} 10 & & & & \\ & 10 & & & \\ & & 8 & & \\ & & & 6 & \\ & & & & 6 \end{bmatrix}$$

To obtain the two columns of the force-transformation matrix, we must determine the bar forces due to unit loads W_1 and W_2 as indicated in Fig. 15.8c. This leads to

$$\begin{bmatrix} q_1 \\ q_2 \\ q_3 \\ q_4 \\ q_5 \end{bmatrix} = \begin{bmatrix} 0.833 & -0.625 \\ -0.833 & -0.625 \\ 0 & 1.0 \\ 0.5 & 0.375 \\ 0.5 & 0.375 \end{bmatrix}\begin{bmatrix} W_1 \\ W_2 \end{bmatrix}$$

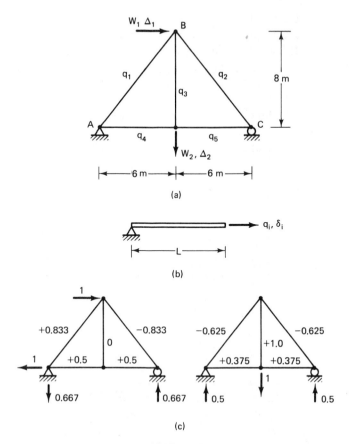

FIGURE 15.8

The structure-flexibility matrix can now be constructed using the relation $[F] = [B]^T[f_c][B]$. Thus we obtain

$$\begin{bmatrix} \Delta_1 \\ \Delta_2 \end{bmatrix} = \frac{1}{AE} \begin{bmatrix} 16.88 & 2.25 \\ 2.25 & 17.50 \end{bmatrix} \begin{bmatrix} W_1 \\ W_2 \end{bmatrix}$$

15.4 ANALYSIS OF INDETERMINATE STRUCTURES

When used to analyze indeterminate structures, the flexibility method is essentially a matrix formulation of the method of consistent deformations. The latter, presented in Chapter 11, can be summarized as follows:

1. The degree of indeterminacy of the structure is ascertained, and a sufficient number of redundants are removed to obtain a determinate structure.
2. The determinate structure is subjected to a set of external forces consisting of the known applied loads and the unknown redundants.

3. The redundants are evaluated by equating the deformations in the determinate structure due to the combined effects of the known loads and the redundants with the corresponding deflections in the indeterminate structure.

4. The member forces, reactions, and structure deformations are obtained using equilibrium and force-deflection relations.

To see how this procedure is carried out using the flexibility method, let us consider the beam in Fig. 15.9a. Choosing the moments at A and C to be the redundants, we obtain for the determinate structure the simply supported beam in Fig. 15.9b. The structure forces, which consist of the known load W_1 and the unknown redundants R_2 and R_3, are shown acting on the determinate structure. The four element forces corresponding to the two elements into which the structure has been subdivided are defined in Fig. 15.9c.

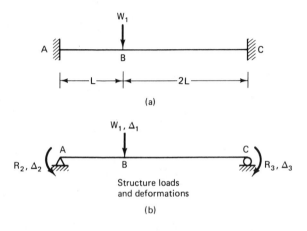

(a)

Structure loads
and deformations

(b)

Element loads
and deformations

(c)   **FIGURE 15.9**

We now proceed to form the structure flexibility matrix $[F]$ for the determinate structure in Fig. 15.9b. The composite element-flexibility matrix is

$$[f_e] = \frac{L}{6EI} \begin{bmatrix} 2 & 1 & 0 & 0 \\ 1 & 2 & 0 & 0 \\ 0 & 0 & 4 & 2 \\ 0 & 0 & 2 & 4 \end{bmatrix}$$

and the force-transformation matrix is given by

$$\begin{bmatrix} q_1 \\ q_2 \\ q_3 \\ q_4 \end{bmatrix} = \left[\begin{array}{c:cc} 0 & -1 & 0 \\ \dfrac{2L}{3} & -\dfrac{2}{3} & -\dfrac{1}{3} \\ \dfrac{2L}{3} & -\dfrac{2}{3} & -\dfrac{1}{3} \\ 0 & 0 & -1 \end{array}\right] \begin{bmatrix} W_1 \\ \hline R_2 \\ R_3 \end{bmatrix} \tag{15.7}$$

As indicated by the dashed line, the force-transformation matrix has been partitioned into two parts, one corresponding to the known load and one corresponding to the unknown redundants.

Next, the structure-flexibility matrix is obtained using the relation $[F] = [B]^{\text{T}} [f_e] [B]$. This leads to

$$\begin{bmatrix} \Delta_1 \\ \hline \Delta_2 \\ \Delta_3 \end{bmatrix} = \frac{L}{18EI} \left[\begin{array}{c:cc} 8L^2 & -10L & -8L \\ \hdashline -10L & 18 & 9 \\ -8L & 9 & 18 \end{array}\right] \begin{bmatrix} W_1 \\ \hline R_2 \\ R_3 \end{bmatrix} \tag{15.8}$$

where F is the 3×3 matrix on the right. The flexibility matrix is partitioned into four submatrices. The horizontal partition divides the unknown deflection Δ_1 from the known deflections Δ_2 and Δ_3, and the vertical partition separates the effects of W_1 from those of R_2 and R_3.

Since Δ_2 and Δ_3, the deflections at the redundants, are equal to zero, we can use the lower part of Eq. (15.8) to solve for the unknown redundants. Thus

$$\begin{bmatrix} 0 \\ 0 \end{bmatrix} = \begin{bmatrix} -10L \\ -8L \end{bmatrix}[W_1] + \begin{bmatrix} 18 & 9 \\ 9 & 18 \end{bmatrix}\begin{bmatrix} R_2 \\ R_3 \end{bmatrix}$$

for which

$$\begin{bmatrix} R_2 \\ R_3 \end{bmatrix} = -\begin{bmatrix} 18 & 9 \\ 9 & 18 \end{bmatrix}^{-1}\begin{bmatrix} -10L \\ -8L \end{bmatrix}[W_1] = \frac{1}{9}\begin{bmatrix} 4L \\ 2L \end{bmatrix}[W_1]$$

Having evaluated the redundants, we can now obtain the member forces and the structure deformations. Substitution of the values for R_2 and R_3 in Eq. (15.7) gives

$$\begin{bmatrix} q_1 \\ q_2 \\ q_3 \\ q_4 \end{bmatrix} = \begin{bmatrix} 0 & -1 & 0 \\ \dfrac{2L}{3} & -\dfrac{2}{3} & -\dfrac{1}{3} \\ \dfrac{2L}{3} & -\dfrac{2}{3} & -\dfrac{1}{3} \\ 0 & 0 & -1 \end{bmatrix} \begin{bmatrix} W_1 \\ \dfrac{4LW_1}{9} \\ \dfrac{2LW_1}{9} \end{bmatrix} = \frac{1}{27}\begin{bmatrix} -12L \\ 8L \\ 8L \\ -6L \end{bmatrix}[W_1]$$

and substitution of the redundants into the upper part of Eq. (15.8) leads to

$$\Delta_1 = \frac{L}{18EI}[8L^2 \quad -10L \quad -8L]\begin{bmatrix} W_1 \\ \dfrac{4LW_1}{9} \\ \dfrac{2LW_1}{9} \end{bmatrix} = \frac{8W_1 L^3}{81EI}$$

The foregoing procedure for analyzing indeterminate structures can be summarized as follows:

1. Decide on the degreee of indeterminacy and remove a sufficient number of redundants to form a determinate structure.
2. To the determinate structure apply the known loads $[W]$ and the unknown redundants $[R]$. These forces together with their corresponding displacements form the structure loads and displacements.
3. Subdivide the structure into elements and define the element forces $[q]$ and the corresponding displacements $[\delta]$.
4. Form the composite element-flexibility matrix $[f_e]$ defined by the relation

$$[\delta] = [f_e][q] \tag{15.9}$$

5. Form the force-transformation matrix $[B]$ for the determinate structure, and partition it into two submatrices corresponding to the known loads $[W]$ and the unknowns $[R]$. Thus

$$[q] = [B_1 \mid B_2]\begin{bmatrix} W \\ -- \\ R \end{bmatrix} \tag{15.10}$$

6. Form the structure flexibility matrix $[F]$ for the determinate structure using the relation $[F] = [B]^T[f_e][B]$, and partition $[F]$ into four submatrices. Thus

$$\begin{bmatrix} \Delta_W \\ ---- \\ \Delta_R \end{bmatrix} = \begin{bmatrix} F_{11} & F_{12} \\ ---- & ---- \\ F_{21} & F_{22} \end{bmatrix}\begin{bmatrix} W \\ --- \\ R \end{bmatrix} \tag{15.11}$$

The horizontal partitioning of $[F]$ corresponds to the separation of $[\Delta]$ into the unknown deflections $[\Delta_W]$ and the known deflections $[\Delta_R]$. Similarly, the vertical partitioning of $[F]$ corresponds to the separation of the loads into $[W]$ and $[R]$.

In the illustrative example it was convenient to obtain $[F]$ as a single matrix and then to partition this matrix. However, if one uses the computer to carry out the calculations, it is desirable to obtain the four submatrices of $[F]$ individually. In view of Eq. (15.10),

$$[F] = [B]^T[f_e][B] = \begin{bmatrix} B_1^T \\ --- \\ B_2^T \end{bmatrix}[f_e][B_1 \mid B_2]$$

and

$$\begin{bmatrix} \Delta_W \\ --- \\ \Delta_R \end{bmatrix} = \begin{bmatrix} B_1^T f_e B_1 & B_1^T f_e B_2 \\ -------- & -------- \\ B_2^T f_e B_1 & B_2^T f_e B_2 \end{bmatrix}\begin{bmatrix} W \\ --- \\ R \end{bmatrix} \tag{15.12}$$

Thus

$$[F_{11}] = [B_1^T f_e B_1]$$
$$[F_{12}] = [B_1^T f_e B_2]$$
$$[F_{21}] = [B_2^T f_e B_1]$$
$$[F_{22}] = [B_2^T f_e B_2]$$

7. Solve the lower part of Eq. (15.11) for $[R]$ using the condition that $[\Delta_R] = [0]$.

$$[0] = [F_{21}][W] + [F_{22}][R]$$

$$[R] = -[F_{22}]^{-1}[F_{21}][W] \tag{15.13}$$

Equation (15.13) gives the solution for the redundants in terms of the known loads $[W]$. To obtain the solution, we must invert $[F_{22}]$, a matrix whose order is equal to the number of redundants.

8. By substituting $[R]$ into Eq. (15.10) we can evaluate the member loads. Thus

$$[q] = [B_1 \mid B_2]\begin{bmatrix} W \\ \hline -F_{22}^{-1}F_{21}W \end{bmatrix}$$

$$[q] = [B_1 - B_2 F_{22}^{-1} F_{21}][W] \tag{15.14}$$

and by substituting $[R]$ into the upper part of Eq. (15.11) we can determine the structure deformations

$$[\Delta_W] = [F_{11} \mid F_{12}]\begin{bmatrix} W \\ \hline -F_{22}^{-1}F_{21}W \end{bmatrix}$$

$$[\Delta_W] = [F_{11} - F_{12}F_{22}^{-1}F_{21}][W] \tag{15.15}$$

Equations (15.14) and (15.15) give the member forces and the structure deformations in terms of the known loads $[W]$. One can think of the matrices to the right of the equality sign in these equations as the force-transformation and flexibility matrices for the indeterminate structures.

Example 15.4

For the frame in Fig. 15.10a find the member forces and the unknown structure deformations.

Let the vertical and horizontal reactions at A be the redundants. The determinate structure obtained by removing the redundants is subjected to three forces consisting of W_1, R_2, and R_3 as shown in Fig. 15.10b. The element forces are defined in Fig. 15.10c.

To form the structure-flexibility matrix requires that we construct $[f_e]$ and $[B]$. Corresponding to the two elements in Fig. 15.10c.

$$[f_e] = \frac{L}{6EI}\begin{bmatrix} 2 & 1 & & \\ 1 & 2 & & \\ \hline & & 4 & 2 \\ & & 2 & 4 \end{bmatrix}$$

and if we use the free bodies in Fig. 15.10d, $[B]$ takes the form

$$\begin{bmatrix} q_1 \\ q_2 \\ q_3 \\ q_4 \end{bmatrix} = \begin{bmatrix} 0 & 0 & 0 \\ 0 & 0 & -L \\ 1 & 0 & -L \\ 1 & 2L & -L \end{bmatrix}\begin{bmatrix} W_1 \\ R_2 \\ R_3 \end{bmatrix} \tag{15.16}$$

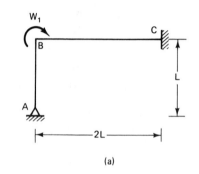

(a)

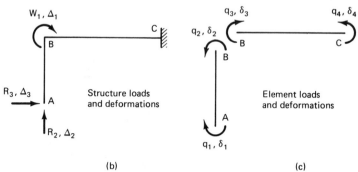

(b)   (c)

FIGURE 15.10

Using the relation $[F] = [B]^T[f_c][B]$, we obtain $[F]$ and write the equation

$$
\begin{bmatrix} \Delta_1 \\ \Delta_2 \\ \Delta_3 \end{bmatrix} = \frac{L}{3EI} \begin{bmatrix} 6 & 6L & -6L \\ 6L & 8L^2 & -6L^2 \\ -6L & -6L^2 & 7L^2 \end{bmatrix} \begin{bmatrix} W_1 \\ R_2 \\ R_3 \end{bmatrix}
$$ (15.17)

Next we evaluate the redundants using the fact that $\Delta_2 = \Delta_3 = 0$. Thus

$$
\begin{bmatrix} \Delta_2 \\ \Delta_3 \end{bmatrix} = \begin{bmatrix} 0 \\ 0 \end{bmatrix} = \begin{bmatrix} 6L \\ -6L \end{bmatrix}[W_1] + \begin{bmatrix} 8L^2 & -6L^2 \\ -6L^2 & 7L^2 \end{bmatrix}\begin{bmatrix} R_2 \\ R_3 \end{bmatrix}
$$

from which

$$
\begin{bmatrix} R_2 \\ R_3 \end{bmatrix} = -\begin{bmatrix} 8L^2 & -6L^2 \\ -6L^2 & 7L^2 \end{bmatrix}^{-1}\begin{bmatrix} 6L \\ -6L \end{bmatrix}[W_1] = \begin{bmatrix} -\dfrac{3W_1}{10L} \\ \dfrac{3W_1}{5L} \end{bmatrix}
$$

Substitution of the redundants into Eq. (15.16) gives the member loads

$$
\begin{bmatrix} q_1 \\ q_2 \\ q_3 \\ q_4 \end{bmatrix} = \begin{bmatrix} 0 & 0 & 0 \\ 0 & 0 & -L \\ 1 & 0 & -L \\ 1 & 2L & -L \end{bmatrix} \begin{bmatrix} 1 \\ -\dfrac{3}{10L} \\ \dfrac{3}{5L} \end{bmatrix}[W_1] = \begin{bmatrix} 0 \\ -\dfrac{6}{10} \\ \dfrac{4}{10} \\ -\dfrac{2}{10} \end{bmatrix}[W_1]
$$

294

The unknown structure deformations are obtained by substituting the redundants into Eq. (15.17).

$$[\Delta_1] = \frac{L}{3EI} \left\{ [6][W_1] + [6L \quad -6L] \begin{bmatrix} -\dfrac{3W_1}{10L} \\[2mm] \dfrac{3W_1}{5L} \end{bmatrix} \right\} = \frac{W_1 L}{5EI}$$

Example 15.5

For the structure in Fig. 15.11a, find the member forces and the unknown structure deformations.

If we let the force in cable CD be the redundant, the determinate structure subjected to the structure loads is as shown in Fig. 15.11b. The element forces are defined in Fig. 15.11c.

To form $[F]$, we need $[f_c]$ and $[B]$. Thus

$$[f_c] = \frac{L}{6EI} \begin{bmatrix} 2 & 1 & & & \\ 1 & 2 & & & \\ \hline & & 4 & 2 & \\ & & 2 & 4 & \\ \hline & & & & \dfrac{6I}{A} \end{bmatrix}$$

and

$$\begin{bmatrix} q_1 \\ q_2 \\ q_3 \\ q_4 \\ q_5 \end{bmatrix} = \begin{bmatrix} -1 & L & 2L \\ -1 & 0 & 2L \\ 0 & 0 & 2L \\ 0 & 0 & 0 \\ 0 & 0 & 1 \end{bmatrix} \begin{bmatrix} W_1 \\ W_2 \\ \hline R_3 \end{bmatrix} \tag{15.18}$$

Using the relation $[F] = [B]^{\mathrm{T}}[f_c][B]$, we obtain

$$\begin{bmatrix} \Delta_1 \\ \Delta_2 \\ \Delta_3 \end{bmatrix} = \frac{L}{6EI} \begin{bmatrix} 6 & -3L & -12L \\ -3L & 2L^2 & 6L^2 \\ \hline -12L & 6L^2 & 40L^2 + \dfrac{6I}{A} \end{bmatrix} \begin{bmatrix} W_1 \\ W_2 \\ \hline R_3 \end{bmatrix} \tag{15.19}$$

Next we determine the redundant from

$$[\Delta_3] = [0] = [-12L \quad 6L^2] \begin{bmatrix} W_1 \\ W_2 \end{bmatrix} + \left[40L^2 + \frac{6I}{A} \right][R_3]$$

which leads to

$$R_3 = \frac{12LW_1 - 6L^2 W_2}{40L^2 + 6I/A}$$

Using the numerical values of W_1, W_2, L, A, and I, we obtain

$$R_3 = 2.31 \text{ k}$$

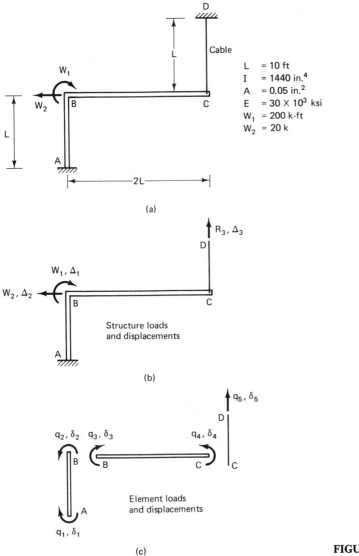

$$
\begin{bmatrix} q_1 \\ q_2 \\ q_3 \\ q_4 \\ q_5 \end{bmatrix} = \begin{bmatrix} -1 & 10 & 20 \\ -1 & 0 & 20 \\ 0 & 0 & 20 \\ 0 & 0 & 0 \\ 0 & 0 & 1 \end{bmatrix} \begin{bmatrix} 200 \\ 20 \\ 2.31 \end{bmatrix} = \begin{bmatrix} 46.2 \text{ k-ft} \\ -153.8 \text{ k-ft} \\ 46.2 \text{ k-ft} \\ 0 \\ 2.31 \text{ k} \end{bmatrix}
$$

(b) — Structure loads and displacements; (c) — Element loads and displacements

FIGURE 15.11

Substitution of the redundant into Eq. (15.18) gives

The unknown structure deformations are obtained by substituting the redundant into

Eq. (15.19). Thus

$$\begin{bmatrix} \Delta_1 \\ \Delta_2 \end{bmatrix} = \frac{10(144)}{6(30 \times 10^3)(1440)} \left\{ \begin{bmatrix} 6 & -30 \\ -30 & 200 \end{bmatrix} \begin{bmatrix} 200 \\ 20 \end{bmatrix} + \begin{bmatrix} -120 \\ 600 \end{bmatrix} [2.31] \right\}$$

$$= \begin{bmatrix} 0.0018 \text{ rad} \\ -0.0034 \text{ ft} \end{bmatrix}$$

PROBLEMS

15.1 and 15.2. Construct the flexibility matrix $[F]$ without subdividing the structure into elements. $EI = $ constant

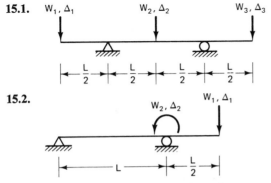

15.1.

15.2.

15.3 to 15.7. Subdivide the structure into elements as indicated, and use the transformation $[F] = [B]^T[f_e][B]$ to form $[F]$. $EI = $ constant

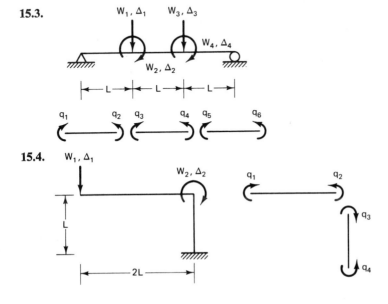

15.3.

15.4.

15.5.

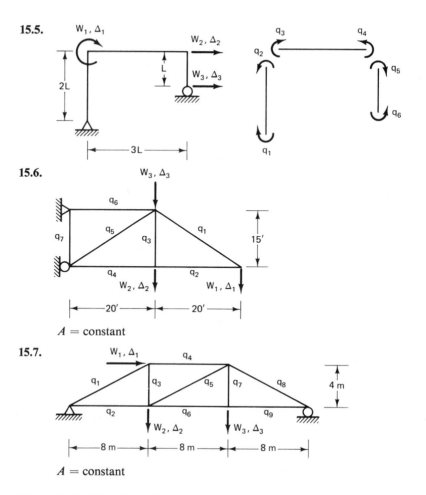

15.6.

$A = \text{constant}$

15.7.

$A = \text{constant}$

15.8 to 15.14. Using the indicated element and structure forces, solve for the redundants, the member forces, and the unknown structure deformations.

15.8. Give answers in terms of L, W_1, W_2, and EI.

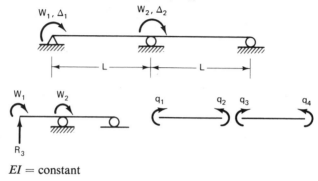

$EI = \text{constant}$

298

15.9.

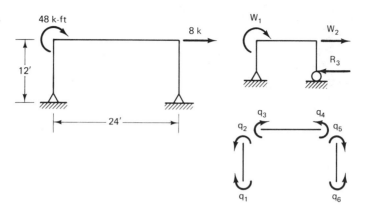

$E = 30 \times 10^3$ ksi
$I = 72$ in^4

15.10. Give answers in terms of L, W_1, and EI.

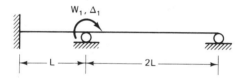

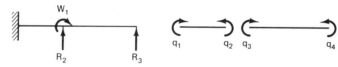

$EI = $ constant

15.11.

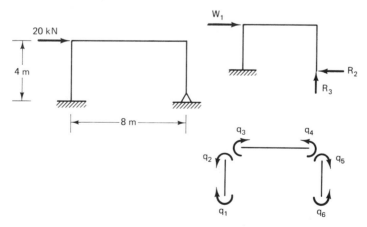

$E = 200 \times 10^6$ kN/m^2
$I = 100 \times 10^6$ mm^4

15.12.

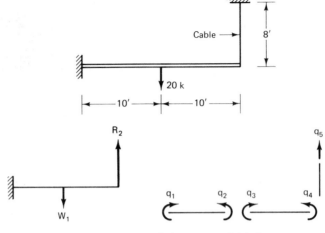

$E = 30 \times 10^3$ ksi, $I_{\text{beam}} = 72$ in^4, $A_{\text{cable}} = 0.1$ in^2

15.13.

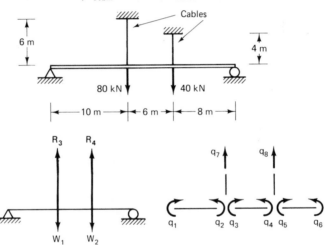

$E = 200 \times 10^6$ kN/m^2, $I_{\text{beam}} = 100 \times 10^6$ mm^4, $A_{\text{cables}} = 20$ mm^2

15.14.

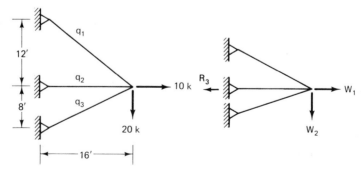

$E = 30 \times 10^3$ ksi, $A = \text{constant} = 1$ in^2

16

Matrix
Stiffness Method

Hawk Falls Bridge, Pa.
(*Courtesy of Bethlehem
Steel Corporation.*)

16.1 INTRODUCTION

The stiffness method, like the flexibility method described in the previous chapter, is a matrix method for analyzing complex structures with the aid of an electronic computer. Whereas the flexibility method uses forces as the independent variables in formulating the load-deformation characteristics of a system, the stiffness method uses joint deformations as the independent variables. The stiffness method is consequently also referred to as the deformation method.

Like the slope-deflection and moment-distribution method, the stiffness method does not involve the concepts of indeterminacy and redundants. As a consequence, it is far easier to program the stiffness method than the flexibility method, and the former is the matrix method used almost exclusively today.

16.2 STIFFNESS MATRIX

In the stiffness method, the forces and deformations in a structure are related to one another by means of stiffness influence coefficients. For example, the moment W_1 acting on the beam in Fig. 16.1a is related to the rotation Δ_1 by the expression

$$W_1 = K_{11} \Delta_1$$

The term K_{11} is a *stiffness influence coefficient.* It is defined as the moment at 1 due to a unit rotation at 1.

If instead of allowing the beam to rotate only at point 1 we also permit it to rotate at point 2, as indicated in Fig. 16.1b, the moment at 1 can be expressed as a function of both the rotations at 1 and at 2. Thus

$$W_1 = K_{11} \Delta_1 + K_{12} \Delta_2$$

In a similar manner we can write an expression for the moment at 2.

$$W_2 = K_{21} \Delta_1 + K_{22} \Delta_2$$

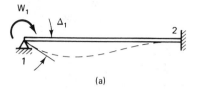

(a)

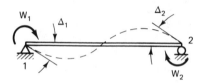

(b)

FIGURE 16.1

Rewriting these equations in matrix form, we obtain

$$\begin{bmatrix} W_1 \\ W_2 \end{bmatrix} = \begin{bmatrix} K_{11} & K_{12} \\ K_{21} & K_{22} \end{bmatrix} \begin{bmatrix} \Delta_1 \\ \Delta_2 \end{bmatrix}$$

or simply

$$[W] = [K][\Delta] \tag{16.1}$$

The matrix $[K]$ that contains the stiffness influence coefficients which relate the forces $[W]$ to the deformations $[\Delta]$ is referred to as the *stiffness matrix*. As pointed out in the previous chapter, it is customary to use the term "force" and the symbol W to refer to moments as well as forces and to designate both rotations and deflections by the symbol Δ.

If both sides of Eq. (16.1) are premultiplied by $[K]^{-1}$, we obtain

$$[\Delta] = [K]^{-1}[W]$$

Comparison of this expression with the equation $[\Delta] = [F][W]$ employed in Chapter 15 indicates that

$$[F] = [K]^{-1}$$

In other words, the flexibility matrix of a structure is the inverse of the stiffness matrix and vice versa.

16.3 ELEMENT STIFFNESS MATRIX

In the stiffness method, as in the flexibility method, the load-deformation characteristics of a structure are obtained from the load-deformation characteristics of a group of elements into which the structure has been subdivided. In other words, we will form the stiffness matrix for a structure out of the stiffness matrices of the individual elements that make up the structure. With this aim in mind, we begin by developing element-stiffness matrices for axially loaded and flexural members.

Axially Loaded Element

Let us construct the stiffness matrix for the member shown in Fig. 16.2a. The vectors in the figure define the forces q_1 and q_2 and the corresponding displacements δ_1 and δ_2 at the ends of the member. They also define the positive directions of these quantities. We refer to these vectors as *element forces* and *displacements*.

The matrix equation that relates the element forces $[q]$ to the corresponding element displacements $[\delta]$ is of the form

$$\begin{bmatrix} q_1 \\ q_2 \end{bmatrix} = \begin{bmatrix} k_{11} & k_{12} \\ k_{21} & k_{22} \end{bmatrix} \begin{bmatrix} \delta_1 \\ \delta_2 \end{bmatrix}$$

To obtain the influence coefficients in the first column of the stiffness matrix, we note that these terms are the forces q_1 and q_2 due to a unit δ_1, with δ_2 equal to zero. Thus k_{11} and k_{21} are the axial forces corresponding to a unit contraction at the left end of

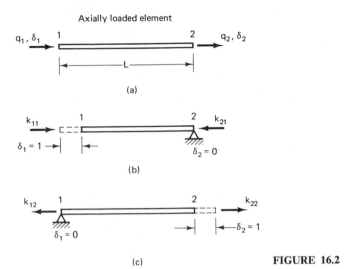

Axially loaded element

(a)

(b)

(c)

FIGURE 16.2

the member, as shown in Fig. 16.2b. These forces have the values

$$k_{11} = \frac{EA}{L}, \qquad k_{21} = -\frac{EA}{L}$$

According to the sign convention defined in Fig. 16.2a, q_1 and q_2 are positive when they act to the right. Thus k_{11} is positive and k_{21} is negative.

Similarly, the second column of the stiffness matrix is obtained by setting $\delta_2 = 1$ and $\delta_1 = 0$. As indicated in Fig. 16.2c, the forces corresponding to these displacements are

$$k_{12} = -\frac{EA}{L}, \qquad k_{22} = \frac{EA}{L}$$

Combining the above results gives

$$\begin{bmatrix} q_1 \\ q_2 \end{bmatrix} = \frac{EA}{L} \begin{bmatrix} 1 & -1 \\ -1 & 1 \end{bmatrix} \begin{bmatrix} \delta_1 \\ \delta_2 \end{bmatrix} \tag{16.2}$$

Equation (16.2) defines the stiffness matrix for an axially loaded element of constant cross-sectional area.

Flexural Element

Let us now construct the stiffness matrix for the flexural element in Fig. 16.3a. The four forces and the corresponding displacements that will be used to describe the behavior of the element are defined in the figure. They include the moments, the shears, and the corresponding rotations and translations at the ends of the member.

The matrix equation that relates these forces and displacements to one another is of the form

$$\begin{bmatrix} q_1 \\ q_2 \\ q_3 \\ q_4 \end{bmatrix} = \begin{bmatrix} k_{11} & k_{12} & k_{13} & k_{14} \\ k_{21} & k_{22} & k_{23} & k_{24} \\ k_{31} & k_{32} & k_{33} & k_{34} \\ k_{41} & k_{42} & k_{43} & k_{44} \end{bmatrix} \begin{bmatrix} \delta_1 \\ \delta_2 \\ \delta_3 \\ \delta_4 \end{bmatrix}$$

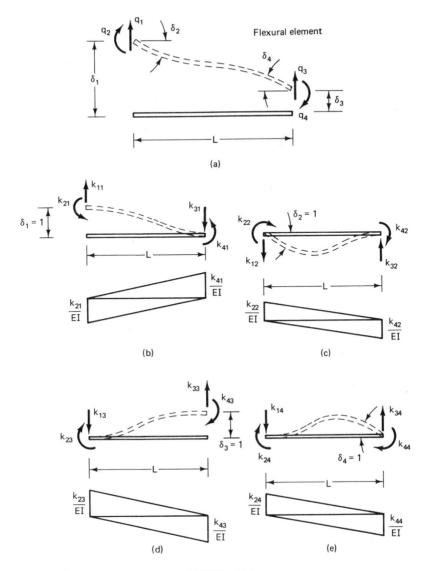

FIGURE 16.3

We begin by determining the influence coefficients in the first column of the stiffness matrix. These terms consist of the element forces q_1 through q_4 that result when $\delta_1 = 1$ and $\delta_2 = \delta_3 = \delta_4 = 0$. Thus we impose a unit vertical displacement at the left end of the member while preventing translation at the right end and rotation at both ends, as shown in Fig. 16.3b. The four member forces that correspond to this deformation can be obtained using the moment-area method.

Since the change in slope between the two ends of the member is zero, the area of the M/EI diagram between these point must vanish. Thus

$$\frac{k_{41}L}{2EI} - \frac{k_{21}L}{2EI} = 0$$

and

$$k_{21} = k_{41} \qquad\qquad (16.3)$$

Furthermore, the moment of the M/EI diagram about the left end of the member is equal to unity. Hence

$$\frac{k_{41}L}{2EI}\left(\frac{2L}{3}\right) - \frac{k_{21}L}{2EI}\left(\frac{L}{3}\right) = 1$$

and in view of Eq. (16.3),

$$k_{41} = k_{21} = \frac{6EI}{L^2}$$

Finally, moment equilibrium of the member about the right end leads to

$$k_{11} = \frac{k_{21} + k_{41}}{L} = \frac{12EI}{L^3}$$

and from equilibrium in the vertical direction we obtain

$$k_{31} = k_{11} = \frac{12EI}{L^3}$$

The forces obtained in the foregoing calculations act in the directions indicated in Fig. 16.3b. To obtain the correct signs for the influence coefficients corresponding to these forces, one must compare the forces with the positive directions defined in Fig. 16.3a. Thus

$$k_{11} = \frac{12EI}{L^3}, \qquad k_{21} = -\frac{6EI}{L^2}, \qquad k_{31} = -\frac{12EI}{L^3}, \qquad k_{41} = -\frac{6EI}{L^2}$$

To obtain the second column of the stiffness matrix, we let $\delta_2 = 1$ and set the remaining three displacements equal to zero as indicated in Fig. 16.3c. For this case the area of the M/EI diagram between the ends of the member is equal to unity, so that

$$\frac{k_{22}L}{2EI} - \frac{k_{42}L}{2EI} = 1$$

and the moment of the M/EI diagram about the left end is zero, so that

$$\frac{k_{22}L}{2EI}\left(\frac{L}{3}\right) - \frac{k_{42}L}{2EI}\left(\frac{2L}{3}\right) = 0$$

Combining these relations gives

$$k_{22} = \frac{4EI}{L}, \qquad k_{42} = \frac{2EI}{L}$$

From vertical equilibrium of the member,

$$k_{12} = k_{32}$$

and moment equilibrium about the right end of the member leads to

$$k_{12} = \frac{k_{22} + k_{42}}{L} = \frac{6EI}{L^2}$$

Comparison of the forces in Fig. 16.3c with the positive directions defined in Fig.

16.3a indicates that all the influence coefficients except k_{12} are positive. Thus

$$k_{12} = -\frac{6EI}{L^2}, \qquad k_{22} = \frac{4EI}{L}, \qquad k_{32} = \frac{6EI}{L^2}, \qquad k_{42} = \frac{2EI}{L}$$

By calculations similar to those above, the influence coefficients for the third and fourth columns can be obtained using Figs. 16.3d and 16.3e. The results of these calculations can then be combined with the preceding ones to give the following element-stiffness matrix:

$$\begin{bmatrix} q_1 \\ q_2 \\ q_3 \\ q_4 \end{bmatrix} = EI \begin{bmatrix} \dfrac{12}{L^3} & -\dfrac{6}{L^2} & -\dfrac{12}{L^3} & -\dfrac{6}{L^2} \\ -\dfrac{6}{L^2} & \dfrac{4}{L} & \dfrac{6}{L^2} & \dfrac{2}{L} \\ -\dfrac{12}{L^3} & \dfrac{6}{L^2} & \dfrac{12}{L^3} & \dfrac{6}{L^2} \\ -\dfrac{6}{L^2} & \dfrac{2}{L} & \dfrac{6}{L^2} & \dfrac{4}{L} \end{bmatrix} \begin{bmatrix} \delta_1 \\ \delta_2 \\ \delta_3 \\ \delta_4 \end{bmatrix} \tag{16.4}$$

Equation (16.4) defines the element-stiffness matrix for a flexural member with constant stiffness EI.

In some flexural structures, such as continuous beams, the ends of an individual member can rotate but are not able to translate. As a consequence, the translations δ_1 and δ_3 together with their corresponding forces q_1 and q_3 can be removed from the element-stiffness matrix. The resulting matrix takes the form

$$\begin{bmatrix} q_2 \\ q_4 \end{bmatrix} = \frac{2EI}{L} \begin{bmatrix} 2 & 1 \\ 1 & 2 \end{bmatrix} \begin{bmatrix} \delta_2 \\ \delta_4 \end{bmatrix} \tag{16.5}$$

16.4 FORMATION OF THE STRUCTURE-STIFFNESS MATRIX FROM ELEMENT-STIFFNESS MATRICES

Having constructed element-stiffness matrices for axially loaded and flexural members, we are now ready to consider the problem of forming the stiffness matrix of an entire structure out of the stiffness matrices of the individual elements into which the structure has been subdivided. To see how this transformation is carried out, let us consider the truss in Fig. 16.4a. We subdivide the structure into two elements and define for each element displacements and forces as indicated in Fig. 16.4b. These forces and the corresponding displacements are related to one another by matrices of the form given in Eq. (16.2). Thus we can write

$$\begin{bmatrix} q_1 \\ q_2 \\ q_3 \\ q_4 \end{bmatrix} = \frac{AE}{L} \begin{bmatrix} 1 & -1 & 0 & 0 \\ -1 & 1 & 0 & 0 \\ 0 & 0 & \dfrac{1}{\sqrt{2}} & -\dfrac{1}{\sqrt{2}} \\ 0 & 0 & -\dfrac{1}{\sqrt{2}} & \dfrac{1}{\sqrt{2}} \end{bmatrix} \begin{bmatrix} \delta_1 \\ \delta_2 \\ \delta_3 \\ \delta_4 \end{bmatrix} \tag{16.6}$$

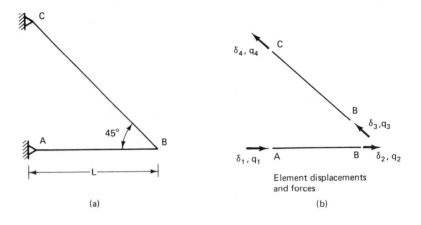

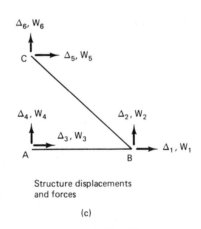

FIGURE 16.4

or

$$[q] = [k_c][\delta] \tag{16.7}$$

where $[k_c]$, the *composite element-stiffness matrix*, contains a basic element-stiffness matrix for each member of the truss.

Next we define six structure deformations and the corresponding structure loads, as shown in Fig. 16.4c. It will be shown later in the chapter that it is advantageous, when numbering the structure deformations, to start with the unknown deformations, as was done here. The matrix equation that relates the structure forces and deformations to one another takes the form

$$[W] = [K][\Delta] \tag{16.8}$$

It is our aim to derive an equation for transforming $[k_c]$, which describes the load-deformation characteristics of the elements taken one at a time, into $[K]$, which describes the load-deformation characteristics of the entire structure. To this end, we

308

will make use of the conditions of compatibility that exist between element and structure deformations.

The element deformations are related to the structure deformations by the matrix equation

$$
\begin{bmatrix} \delta_1 \\ \delta_2 \\ \delta_3 \\ \delta_4 \end{bmatrix} =
\begin{bmatrix}
0 & 0 & 1 & 0 & 0 & 0 \\
1 & 0 & 0 & 0 & 0 & 0 \\
-\cos\alpha & \sin\alpha & 0 & 0 & 0 & 0 \\
0 & 0 & 0 & 0 & -\cos\alpha & \sin\alpha
\end{bmatrix}
\begin{bmatrix} \Delta_1 \\ \Delta_2 \\ \Delta_3 \\ \Delta_4 \\ \Delta_5 \\ \Delta_6 \end{bmatrix}
\tag{16.9}
$$

or

$$
[\delta] = [T][\Delta] \tag{16.10}
$$

The matrix $[T]$ is called the *deformation-transformation matrix*. Its columns are obtained by applying unit values of the structure deformations Δ_1 through Δ_6, one at a time, and determining the corresponding element deformations. For example, the first column in $[T]$ is obtained by letting $\Delta_1 = 1$ and setting Δ_2 through Δ_6 equal to zero, as shown in Fig. 16.5a. The horizontal displacement of joint B results in $\delta_2 = 1$ and $\delta_3 = -\cos\alpha$. The negative sign of δ_3 is due to the fact that a positive Δ_1 causes a negative δ_3. Since joints A and C do not move when Δ_1 is applied, δ_1 and δ_4 are equal to zero.

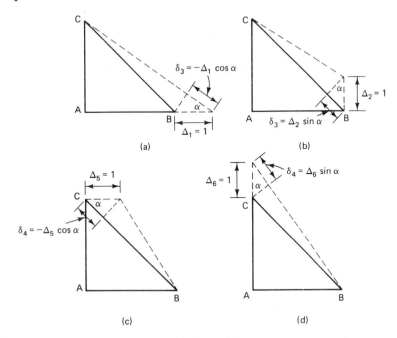

FIGURE 16.5

With the aid of Figs. 16.5b, 16.5c, and 16.5d, the remaining columns in $[T]$ can be constructed. It should be noted that as a consequence of the assumption of small deformations, structure deformations normal to a member are assumed to cause zero deformation in that member.

To complete the derivation of the equation that transforms $[k_c]$ into $[K]$, we make use of the principle of conservation of energy. The work performed by the structure forces acting on the entire structure must be equal to the work performed by the element forces acting on all the elements into which the structure has been subdivided. Thus

$$\frac{1}{2}[\Delta]^{\mathrm{T}}[W] = \frac{1}{2}[\delta]^{\mathrm{T}}[q]$$

Substitution of Eqs. (16.8) and (16.7) for $[W]$ and $[q]$ into the above relation gives

$$[\Delta]^{\mathrm{T}}[K][\Delta] = [\delta]^{\mathrm{T}}[k_c][\delta]$$

and in view of Eq. (16.10) we obtain

$$[\Delta]^{\mathrm{T}}[K][\Delta] = [\Delta]^{\mathrm{T}}[T]^{\mathrm{T}}[k_c][T][\Delta]$$

which reduces to

$$[K] = [T]^{\mathrm{T}}[k_c][T] \tag{16.11}$$

The structure-stiffness matrix $[K]$ can thus be obtained from the composite element-stiffness matrix $[k_c]$ if the latter is premultiplied by $[T]^{\mathrm{T}}$ and postmultiplied by $[T]$. Equation (16.11) is very similar to Eq. (15.6), the corresponding expression in the flexibility method. The only difference is that the transformation from element behavior to structure behavior is carried out using a force-transformation matrix $[B]$ in the flexibility method, whereas a deformation-transformation matrix $[T]$ is used in the stiffness method.

If the operations indicated by Eq. (16.11) are carried out using $[k_c]$ given by Eq. (16.6) and $[T]$ given by Eq. (16.9), one obtains

$$[K] = \frac{AE}{4L}\begin{bmatrix} 4+\sqrt{2} & -\sqrt{2} & -4 & 0 & -\sqrt{2} & \sqrt{2} \\ -\sqrt{2} & \sqrt{2} & 0 & 0 & \sqrt{2} & -\sqrt{2} \\ -4 & 0 & 4 & 0 & 0 & 0 \\ 0 & 0 & 0 & 0 & 0 & 0 \\ -\sqrt{2} & \sqrt{2} & 0 & 0 & \sqrt{2} & -\sqrt{2} \\ \sqrt{2} & -\sqrt{2} & 0 & 0 & -\sqrt{2} & \sqrt{2} \end{bmatrix} \tag{16.12}$$

This is the structure-stiffness matrix for the truss in Fig. 16.4.

16.5 THE DIRECT STIFFNESS METHOD OF FORMING THE STRUCTURE-STIFFNESS MATRIX

The procedure presented in the previous section, for obtaining the structure-stiffness matrix, is very similar to the one used in the flexibility method. In both instances, the force-deflection relationship of an entire structure is obtained from a matrix containing the force-deflection relationships of all the individual elements into which the structure has been subdivided. The disadvantage of this procedure is that it does not readily lend itself to automatic computation. As a consequence, we will now introduce a slightly different method for obtaining the structure-stiffness matrix, one that can be entirely automated. This alternative procedure, which will be used from here on, is called the *direct stiffness method.*

In the stiffness method, deformations are used as the independent variables and the behavior of the structure is formulated in terms of these variables. Accordingly, the deformations are often referred to as coordinates and it is customary to speak of the transformation from the element-stiffness matrices to the structure-stiffness matrix as a transformation of coordinates. In the direct stiffness method, this transformation from element to structure coordinates is carried out separately for each element and the resulting matrices are then combined to form the structure-stiffness matrix. By comparison, the previous method combined all of the element-stiffness matrices into a single composite element-stiffness matrix and then transformed this composite matrix from element to structure coordinates.

In the direct stiffness method the transformation from element to structure coordinates is carried out, as before, using deformation compatibility relations between element and structure deformations. However, whereas a single $[T]$ matrix was used previously for the entire structure, we will now write a separate $[T]$ matrix for each element. In other words, a relation of the form given by Eq. (16.10) is now written for each element. Thus

$$[\delta]_n = [T]_n [\Delta]_n \qquad (16.13)$$

where $[T]_n$ is the matrix that relates the element deformations of element n to the structure deformations at the extremities of that element.

Since the vectors used to represent forces are identical to those used to represent deformations, the element and structure forces corresponding to a given element must be related to each other in exactly the same manner as the element and structure deformations. That is,

$$[q]_n = [T]_n [W]_n \qquad (16.14)$$

where $[q]_n$ contains the element forces for element n and $[W]_n$ contains the structure forces at the extremities of the element.

Using the transformation matrix $[T]_n$, we will now develop the necessary equation for carrying out the transformation from element to structure coordinates for element n. Starting with the relation between forces and deformations for the element, that is,

$$[q]_n = [k]_n [\delta]_n$$

and making use of Eqs. (16.13) and (16.14), we obtain

$$[T]_n [W]_n = [k]_n [T]_n [\Delta]_n$$

or

$$[W]_n = [T]_n^{-1} [k]_n [T]_n [\Delta]_n \qquad (16.15)$$

Since $[T]_n$ represents an orthogonal transformation,

$$[T]_n^{-1} = [T]_n^{\mathrm{T}}$$

and Eq. (16.15) becomes

$$[W]_n = [T]_n^{\mathrm{T}}[k]_n[T]_n[\Delta]_n$$

from which it is evident that

$$[K]_n = [T]_n^{\mathrm{T}}[k]_n[T]_n \qquad (16.16)$$

Equation (16.16) transforms the stiffness matrix for any element n from element to structure coordinates. The similarity between Eqs. (16.16) and (16.11) is obvious. The difference is that Eq. (16.11) transforms a composite element-stiffness matrix, containing matrices for each element, into a stiffness matrix for the entire structure, whereas Eq. (16.16) transforms only a single element-stiffness matrix from element to structure coordinates. As a consequence, the direct stiffness method requires that Eq. (16.16) be applied to each element individually and that the resulting matrices be then combined to form the stiffness matrix for the entire structure.

At first sight the direct stiffness method appears to be more involved than the use of Eq. (16.11). However, this is not the case. In the direct stiffness method it is possible to construct a matrix $[T]_n$ that is applicable to any element. By comparison, the use of Eq. (16.11) requires that $[T]$ be rederived for every new structure considered.

To illustrate the direct stiffness method, let us again make use of the truss in Fig. 16.4. The first step in the procedure is to construct a transformation matrix $[T]_n$ for each member of the truss. This matrix relates the two element deformations of the member to the four structure deformations at the extremities of the member. The individual terms in $[T]_n$ are obtained by applying unit structure deformations, one at a time, and determining the corresponding element deformations. Thus for element ab we obtain

$$\begin{bmatrix} \delta_1 \\ \delta_2 \end{bmatrix} = \begin{bmatrix} 1 & 0 & 0 & 0 \\ 0 & 0 & 1 & 0 \end{bmatrix} \begin{bmatrix} \Delta_3 \\ \Delta_4 \\ \Delta_1 \\ \Delta_2 \end{bmatrix}$$

and for element bc

$$\begin{bmatrix} \delta_3 \\ \delta_4 \end{bmatrix} = \frac{1}{\sqrt{2}} \begin{bmatrix} -1 & 1 & 0 & 0 \\ 0 & 0 & -1 & 1 \end{bmatrix} \begin{bmatrix} \Delta_1 \\ \Delta_2 \\ \Delta_5 \\ \Delta_6 \end{bmatrix}$$

Using the above transformation matrices and carrying out the operations indicated by Eq. (16.16) gives

$$
[K]_{ab} = \begin{array}{c} 3 \\ 4 \\ 1 \\ 2 \end{array} \begin{bmatrix} 1 & 0 \\ 0 & 0 \\ 0 & 1 \\ 0 & 0 \end{bmatrix} \frac{AE}{L} \begin{bmatrix} 1 & -1 \\ -1 & 1 \end{bmatrix} \overset{\begin{array}{cccc} 3 & 4 & 1 & 2 \end{array}}{\begin{bmatrix} 1 & 0 & 0 & 0 \\ 0 & 0 & 1 & 0 \end{bmatrix}}
$$

$$
= \frac{AE}{L} \overset{\begin{array}{cccc} 3 & 4 & 1 & 2 \end{array}}{\begin{bmatrix} 1 & 0 & -1 & 0 \\ 0 & 0 & 0 & 0 \\ -1 & 0 & 1 & 0 \\ 0 & 0 & 0 & 0 \end{bmatrix}} \begin{array}{c} 3 \\ 4 \\ 1 \\ 2 \end{array} \qquad (16.17)
$$

and

$$
[K]_{bc} = \begin{array}{c} 1 \\ 2 \\ 5 \\ 6 \end{array} \frac{1}{\sqrt{2}} \begin{bmatrix} -1 & 0 \\ 1 & 0 \\ 0 & -1 \\ 0 & 1 \end{bmatrix} \frac{AE}{\sqrt{2}\,L} \begin{bmatrix} 1 & -1 \\ -1 & 1 \end{bmatrix} \frac{1}{\sqrt{2}} \overset{\begin{array}{cccc} 1 & 2 & 5 & 6 \end{array}}{\begin{bmatrix} -1 & 1 & 0 & 0 \\ 0 & 0 & -1 & 1 \end{bmatrix}}
$$

$$
= \frac{AE}{2\sqrt{2}\,L} \overset{\begin{array}{cccc} 1 & 2 & 5 & 6 \end{array}}{\begin{bmatrix} 1 & -1 & -1 & 1 \\ -1 & 1 & 1 & -1 \\ -1 & 1 & 1 & -1 \\ 1 & -1 & -1 & 1 \end{bmatrix}} \begin{array}{c} 1 \\ 2 \\ 5 \\ 6 \end{array} \qquad (16.18)
$$

The foregoing operations transformed the stiffness matrices for members *AB* and *BC* from element to structure coordinates. In other words, whereas $[k]_{ab}$ related the element forces for member *AB* to the corresponding element deformations, $[K]_{ab}$ relates the structure forces at the extremities of the element to the corresponding structure deformations. The row and column designations listed to the right and at the top of each $[K]_n$ matrix identify the specific structure forces and deformations that correspond to each element. In accordance with the law of matrix multiplication, the row and column designation of any term in $[K]_n$ can be obtained from the corresponding row in $[T]_n^T$ and the corresponding column in $[T]_n$.

Once the stiffness matrices of the individual members have been transformed from element to structure coordinates, there remains the task of combining these matrices to form the stiffness matrix for the entire structure. This step is carried out by placing the individual terms of the element-stiffness matrices into their proper position in the structure-stiffness matrix. The row and column of $[K]$ into which any term from $[K]_n$ goes is determined by the row and column designation of that term. For example, the four terms in the upper left-hand corner of $[K]_{ab}$ go into rows 3 and

4 and columns 3 and 4 of $[K]$. When more than one element-stiffness term goes into a single slot in the structure-stiffness matrix, the terms are added.

Combining $[K]_{ab}$ and $[K]_{bc}$ in the manner described above leads to the following structure-stiffness matrix:

$$[K] = \frac{AE}{4L}
\begin{array}{c}
 \\
\\
\end{array}
\begin{array}{cccccc}
1 & 2 & 3 & 4 & 5 & 6 \\
\end{array}$$

$$[K] = \frac{AE}{4L}
\begin{bmatrix}
4+\sqrt{2} & -\sqrt{2} & -4 & 0 & -\sqrt{2} & \sqrt{2} \\
-\sqrt{2} & \sqrt{2} & 0 & 0 & \sqrt{2} & -\sqrt{2} \\
-4 & 0 & 4 & 0 & 0 & 0 \\
0 & 0 & 0 & 0 & 0 & 0 \\
-\sqrt{2} & \sqrt{2} & 0 & 0 & \sqrt{2} & -\sqrt{2} \\
\sqrt{2} & -\sqrt{2} & 0 & 0 & -\sqrt{2} & \sqrt{2}
\end{bmatrix}
\begin{array}{c}
1 \\ 2 \\ 3 \\ 4 \\ 5 \\ 6
\end{array} \qquad (16.19)$$

It should be noted that whereas the above procedure for forming $[K]$ is tedious when carried out by hand, the operation becomes relatively simple when performed by a computer.

The physical significance of adding terms of the element-stiffness matrices to obtain terms in the structure-stiffness matrix can be explained as follows. An external load W applied to any joint of a structure must be resisted by all the members framing into that joint. Thus each term in an individual $[K]_n$ matrix represents the resistance of member n to one of the external forces at its ends, and the process of combining terms of $[K]_n$ matrices to obtain terms in $[K]$ is an application of the laws of equilibrium at the joints of the structure.

To avoid confusing the various stiffness matrices introduced in this section, it may help to remember that a lower-case symbol refers to element coordinates and an upper-case symbol to structure coordinates. In addition, a subscript denotes that the matrix describes a single member, whereas the absence of a subscript means that it refers to the entire structure. Thus $[k]_n$ is an element-stiffness matrix in element coordinates and $[K]_n$ is an element-stiffness matrix in structure coordinates. By comparison, $[K]$ is the stiffness matrix for the entire structure.

16.6 ANALYSIS OF TRUSSES BY THE DIRECT STIFFNESS METHOD

Formation of the Structure-Stiffness Matrix

The great advantage of the direct stiffness method is that the formation of the structure-stiffness matrix can be completely automated. In this section we will demonstrate how this is accomplished for a truss. Not only will we construct a transformation matrix $[T]_n$ that is applicable to any axially loaded bar, but we will also form a member-stiffness matrix $[K]_n$, in structure coordinates, that is likewise generally applicable.

Let us consider the truss members in Fig. 16.6. To systematize the formation of the $[K]_n$ matrix for members such as these, we adopt the following procedure:

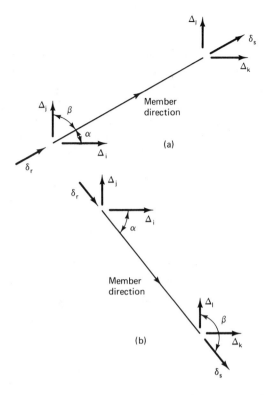

Member direction

(a)

Member direction

(b)

FIGURE 16.6

1. The member is assigned a direction, and δ_r and δ_s, the member deformations at the tail end and front end of the member, are assumed to be in this direction.

2. The x and y structure deformations, at the tail and front end of the member, are given the designations Δ_i, Δ_j, Δ_k, and Δ_l respectively. The positive directions for these vectors are always taken to be to the right and up.

3. The angles between the vectors representing the member deformations and those representing the x and y structure deformations are designated as α and β, and the cosines of these angles by λ and μ. Thus

$$\lambda = \cos \alpha, \qquad \mu = \cos \beta$$

It is customary to refer to λ and μ as the *direction cosines* of the member.

Since a unit structure deformation at one end of a member gives rise to a member deformation at the same end equal to the cosine of the angle between the two deformations, and since structure deformations at one end do not cause element deformations at the other end, the relationship between element and structure deformations can be written in the form

$$\begin{bmatrix} \delta_r \\ \delta_s \end{bmatrix} = \begin{bmatrix} \lambda & \mu & 0 & 0 \\ 0 & 0 & \lambda & \mu \end{bmatrix} \begin{bmatrix} \Delta_i \\ \Delta_j \\ \Delta_k \\ \Delta_l \end{bmatrix} \tag{16.20}$$

Equation (16.20) gives the transformation matrix $[T]_n$ for any axially loaded bar, regardless of the direction of the member deformations relative to the structure deformations. The fact that positive structure deformations give rise to positive or negative element deformations, depending on whether the angle between the two vectors is smaller or larger than 90°, is taken care of in Eq. (16.20) by the signs of λ and μ. For example, the angles α and β are both less than 90° for the bar in Fig. 16.6a. As a consequence, positive structure deformations produce positive member deformations, and this is accounted for in Eq. (16.20) by λ and μ both being positive. By comparison, α is less than 90° and β is greater than 90° for the bar in Fig. 16.6b. Accordingly, positive structure deformations in the x-direction produce positive member deformations, whereas positive structure deformations in the y-direction lead to negative member deformations. Again this is taken care of in Eq. (16.20) by λ being positive and μ negative.

We have now available to us, in general form, both an element-stiffness matrix $[k]_n$ and a transformation matrix $[T]_n$. Using these matrices and the relation $[K]_n = [T]_n^T[k]_n[T]_n$, we can derive a member-stiffness matrix in structure coordinates that is applicable to any axially loaded bar. Thus

$$[k]_n[T]_n = \frac{AE}{L}\begin{bmatrix} 1 & -1 \\ -1 & 1 \end{bmatrix}\begin{matrix} i & j & k & l \\ \begin{bmatrix} \lambda & \mu & 0 & 0 \\ 0 & 0 & \lambda & \mu \end{bmatrix} \end{matrix}$$

$$= \frac{AE}{L}\begin{matrix} i & j & k & l \\ \begin{bmatrix} \lambda & \mu & -\lambda & -\mu \\ -\lambda & -\mu & \lambda & \mu \end{bmatrix} \end{matrix} \qquad (16.21)$$

and

$$[K]_n = [T]_n^T[k]_n[T]_n = \begin{matrix} i \\ j \\ k \\ l \end{matrix}\begin{bmatrix} \lambda & 0 \\ \mu & 0 \\ 0 & \lambda \\ 0 & \mu \end{bmatrix}\frac{AE}{L}\begin{matrix} i & j & k & l \\ \begin{bmatrix} \lambda & \mu & -\lambda & -\mu \\ -\lambda & -\mu & \lambda & \mu \end{bmatrix} \end{matrix}$$

from which

$$[K]_n = \frac{AE}{L}\begin{bmatrix} \lambda^2 & \lambda\mu & -\lambda^2 & -\lambda\mu \\ \lambda\mu & \mu^2 & -\lambda\mu & -\mu^2 \\ -\lambda^2 & -\lambda\mu & \lambda^2 & \lambda\mu \\ -\lambda\mu & -\mu^2 & \lambda\mu & \mu^2 \end{bmatrix}\begin{matrix} i \\ j \\ k \\ l \end{matrix} \qquad (16.22)$$

Equation (16.22) gives the element-stiffness matrix in structure coordinates for any axially loaded bar. To construct $[K]_n$ for a given member, one need only know the values of λ and μ for the member and the structure coordinates i, j, k, and l at its extremities. The values of λ and μ determine those of the terms in $[K]_n$, and i, j, k, and l give the row and column designations of $[K]_n$. The latter are used to locate the individual terms of $[K]_n$ in the stiffness matrix for the entire structure.

Example 16.1

As an illustration of the direct stiffness method, let us employ Eq. (16.22) to construct the structure-stiffness matrix for the truss in Fig. 16.7a.

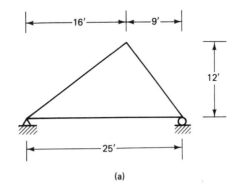

(a)

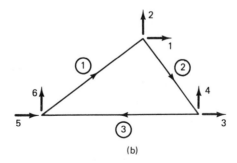

(b)

Member	Length	α	β	λ	μ	i	j	k	l
1	20′	36.87°	53.13°	0.8	0.6	5	6	1	2
2	15′	53.13°	143.13°	0.6	−0.8	1	2	3	4
3	25′	180°	90°	−1.0	0	3	4	5	6

(c)

FIGURE 16.7

We begin by numbering the members, choosing directions for them, and defining the structure coordinates. Next we list in tabular form the data necessary to form the individual $[K]_n$ matrices. Using these data and Eq. (16.22) gives

$$[K]_1 = \frac{AE}{5} \begin{array}{cccc} 5 & 6 & 1 & 2 \\ \begin{bmatrix} 0.16 & 0.12 & -0.16 & -0.12 \\ 0.12 & 0.09 & -0.12 & -0.09 \\ -0.16 & -0.12 & 0.16 & 0.12 \\ -0.12 & -0.09 & 0.12 & 0.09 \end{bmatrix} & \begin{array}{c} 5 \\ 6 \\ 1 \\ 2 \end{array} \end{array}$$

$$[K]_2 = \frac{AE}{5} \begin{array}{cccc} 1 & 2 & 3 & 4 \\ \begin{bmatrix} 0.12 & -0.16 & -0.12 & 0.16 \\ -0.16 & 0.21 & 0.16 & -0.21 \\ -0.12 & 0.16 & 0.12 & -0.16 \\ 0.16 & -0.21 & -0.16 & 0.21 \end{bmatrix} & \begin{array}{c} 1 \\ 2 \\ 3 \\ 4 \end{array} \end{array}$$

and

$$[K]_3 = \frac{AE}{5} \begin{array}{cccc} 3 & 4 & 5 & 6 \\ \begin{bmatrix} 0.20 & 0 & -0.20 & 0 \\ 0 & 0 & 0 & 0 \\ -0.20 & 0 & 0.20 & 0 \\ 0 & 0 & 0 & 0 \end{bmatrix} & \begin{array}{c} 3 \\ 4 \\ 5 \\ 6 \end{array} \end{array}$$

Finally, by placing the individual terms of the above matrices into their correct positions in the structure-stiffness matrix we obtain

$$[K] = \frac{AE}{5} \begin{array}{cccccc} 1 & 2 & 3 & 4 & 5 & 6 \\ \begin{bmatrix} 0.28 & -0.04 & -0.12 & 0.16 & -0.16 & -0.12 \\ -0.04 & 0.30 & 0.16 & -0.21 & -0.12 & -0.09 \\ -0.12 & 0.16 & 0.32 & -0.16 & -0.20 & 0 \\ 0.16 & -0.21 & -0.16 & 0.21 & 0 & 0 \\ -0.16 & -0.12 & -0.20 & 0 & 0.36 & 0.12 \\ -0.12 & -0.09 & 0 & 0 & 0.12 & 0.09 \end{bmatrix} & \begin{array}{c} 1 \\ 2 \\ 3 \\ 4 \\ 5 \\ 6 \end{array} \end{array}$$

Solution for Unknown Displacements and Forces

The stiffness method, like the flexibility method, consists of two parts. In the first, the structure-stiffness matrix is formed out of the stiffness matrices of the individual elements into which the structure has been subdivided, and in the second part, the structure-stiffness matrix is used to determine the deflections, reactions, and internal forces of the structure. We have dealt with the first of these two steps in the preceding pages and are now ready to consider the second.

The equation $[W] = [K][\Delta]$ relates the structure loads to the corresponding structure displacements. In order to solve this equation for the unknown structure displacements and for the reactions, it is useful to partition each of the matrices involved. Thus we write

$$\begin{bmatrix} W_k \\ \hline W_u \end{bmatrix} = \begin{bmatrix} K_{11} & K_{12} \\ \hline K_{21} & K_{22} \end{bmatrix} \begin{bmatrix} \Delta_u \\ \hline \Delta_k \end{bmatrix} \tag{16.23}$$

As indicated, the structure-deformation matrix $[\Delta]$ is partitioned into two submatrices $[\Delta_u]$ and $[\Delta_k]$ corresponding to the unknown joint displacements and those prescribed by the boundary conditions. The load matrix $[W]$ is likewise partitioned into two submatrices $[W_k]$ and $[W_u]$. The first of these, $[W_k]$, contains the known loads applied at the joints free to move, whereas the second, $[W_u]$, is made up of the unknown reactions acting at the joints restrained against motion. Thus $[W_k]$ and $[W_u]$ have the same number of terms as $[\Delta_u]$ and $[\Delta_k]$, respectively. The stiffness matrix $[K]$ is then divided into four submatrices in accordance with the partitioning of $[\Delta]$ and $[W]$.

In most instances $[\Delta_k]$ will contain only zero terms, and Eq. (16.23) can accordingly be separated into the two relations

$$[W_k] = [K_{11}][\Delta_u] \tag{16.24}$$

and

$$[W_u] = [K_{21}][\Delta_u] \tag{16.25}$$

Multiplying both sides of Eq. (16.24) by $[K_{11}]^{-1}$ gives

$$[\Delta_u] = [K_{11}]^{-1}[W_k] \tag{16.26}$$

from which the unknown nodal displacements can be obtained. Once these deformations are known, the reactions $[W_u]$ can be found using Eq. (16.25). The reason for separating the unknown from the known joint displacements, when assigning numbers to the structure degrees of freedom, should now be obvious.

The final step in the solution involves the calculation of the member loads $[q]$. Starting with the relation between element loads and displacements

$$[q]_n = [k]_n[\delta]_n$$

and making use of the equation

$$[\delta]_n = [T]_n[\Delta]_n$$

we obtain

$$[q]_n = [k]_n[T]_n[\Delta]_n = [kT]_n[\Delta]_n \tag{16.27}$$

which, in view of Eq. (16.21), can be written as

$$\begin{bmatrix} q_r \\ q_s \end{bmatrix} = \frac{AE}{L} \begin{bmatrix} \lambda & \mu & -\lambda & -\mu \\ -\lambda & -\mu & \lambda & \mu \end{bmatrix} \begin{bmatrix} \Delta_i \\ \Delta_j \\ \Delta_k \\ \Delta_l \end{bmatrix} \tag{16.28}$$

Since q_r is always numerically equal to q_s, it is only necessary to solve for one of these forces. Thus

$$[q_s] = \frac{AE}{L}[-\lambda \quad -\mu \quad \lambda \quad \mu] \begin{bmatrix} \Delta_i \\ \Delta_j \\ \Delta_k \\ \Delta_l \end{bmatrix} \tag{16.29}$$

Because q_s was assumed to be a tension force, a positive answer means that the bar is in tension.

Example 16.2

Determine the reactions and the bar forces for the truss in Fig. 16.8a.

We begin by numbering the members, choosing directions for them, and defining the structure coordinates, as shown in Fig. 16.8b. Since we intend to evaluate only one force per member, namely, the one at the head of the member, it is convenient to give this force the same number that we have assigned to the member.

Next we list in tabular form, as shown in Fig. 16.8c, the data necessary for forming the individual element stiffness matrices. It is essential to remember that i, j, k, and l refer to the x and y structure coordinates at the tail and head of the member, respectively.

From the data in Fig. 16.8c and Eq. (16.22) the element stiffness matrices are constructed.

$$[K]_1 = \frac{AE}{5} \begin{bmatrix} 0.09 & -0.12 & -0.09 & 0.12 \\ -0.12 & 0.16 & 0.12 & -0.16 \\ -0.09 & 0.12 & 0.09 & -0.12 \\ 0.12 & -0.16 & -0.12 & 0.16 \end{bmatrix} \begin{matrix} 3 \\ 4 \\ 1 \\ 2 \end{matrix}$$

with column headings $\begin{matrix} 3 & 4 & 1 & 2 \end{matrix}$

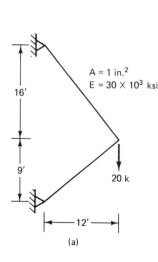

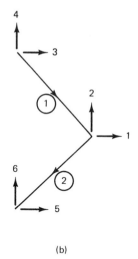

(a)   (b)

Member	Length	α	β	λ	μ	i	j	k	l
1	20'	53.13°	143.13°	0.6	−0.8	3	4	1	2
2	15'	143.13°	126.87°	−0.8	−0.6	1	2	5	6

(c)

FIGURE 16.8

$$[K]_2 = \frac{AE}{5}
\begin{array}{c}
1 \quad\quad 2 \quad\quad 5 \quad\quad 6 \\
\begin{bmatrix}
0.21 & 0.16 & -0.21 & -0.16 \\
0.16 & 0.12 & -0.16 & -0.12 \\
-0.21 & -0.16 & 0.21 & 0.16 \\
-0.16 & -0.12 & 0.16 & 0.12
\end{bmatrix}
\begin{array}{c} 1 \\ 2 \\ 5 \\ 6 \end{array}
\end{array}$$

Combining the element-stiffness matrices, we obtain the structure-stiffness matrix.

$$\begin{bmatrix} W_1 \\ W_2 \\ W_3 \\ W_4 \\ W_5 \\ W_6 \end{bmatrix} = \frac{AE}{5}
\begin{bmatrix}
0.30 & 0.04 & -0.09 & 0.12 & -0.21 & -0.16 \\
0.04 & 0.28 & 0.12 & -0.16 & -0.16 & -0.12 \\
-0.09 & 0.12 & 0.09 & -0.12 & 0 & 0 \\
0.12 & -0.16 & -0.12 & 0.16 & 0 & 0 \\
-0.21 & -0.16 & 0 & 0 & 0.21 & 0.16 \\
-0.16 & -0.12 & 0 & 0 & 0.16 & 0.12
\end{bmatrix}
\begin{bmatrix} \Delta_1 \\ \Delta_2 \\ \Delta_3 \\ \Delta_4 \\ \Delta_5 \\ \Delta_6 \end{bmatrix}$$

Next we subdivide $[\Delta]$ and $[W]$ into unknown and known deformations and loads and partition $[K]$ accordingly.

Using Eq. (16.26), we are now ready to solve for the unknown structure deformations Δ_1 and Δ_2.

$$\begin{bmatrix} \Delta_1 \\ \Delta_2 \end{bmatrix} = \frac{5}{AE}\begin{bmatrix} 0.30 & 0.04 \\ 0.04 & 0.28 \end{bmatrix}^{-1}\begin{bmatrix} 0 \\ -20 \end{bmatrix} = \frac{1}{AE}\begin{bmatrix} 48.54 \\ -364.08 \end{bmatrix}$$

The reactions are determined using Eq. (16.25).

$$\begin{bmatrix} W_3 \\ W_4 \\ W_5 \\ W_6 \end{bmatrix} = \frac{AE}{5}\begin{bmatrix} -0.09 & 0.12 \\ 0.12 & -0.16 \\ -0.21 & -0.16 \\ -0.16 & -0.12 \end{bmatrix}\frac{1}{AE}\begin{bmatrix} 48.54 \\ -364.08 \end{bmatrix} = \begin{bmatrix} -9.61 \\ +12.82 \\ +9.61 \\ +7.18 \end{bmatrix}$$

Finally, we calculate the bar forces using Eq. (16.29).

$$q_1 = \frac{AE}{20}\begin{array}{c} 3 \quad\ 4 \quad\ 1 \quad\ 2 \\ [-0.6 \quad 0.8 \quad 0.6 \quad -0.8] \end{array}\frac{1}{AE}\begin{bmatrix} 0 \\ 0 \\ 48.54 \\ -364.08 \end{bmatrix} = 16.0$$

$$q_2 = \frac{AE}{15}\begin{array}{c} 1 \quad\ 2 \quad\ 5 \quad\ 6 \\ [0.8 \quad 0.6 \quad -0.8 \quad -0.6] \end{array}\frac{1}{AE}\begin{bmatrix} 48.54 \\ -364.08 \\ 0 \\ 0 \end{bmatrix} = -12.0$$

An alternative way of evaluating the bar forces, which is especially convenient when using the computer, is to combine the above matrices and eliminate the zero terms.

Thus

$$
\begin{bmatrix} q_1 \\ q_2 \end{bmatrix} = \begin{bmatrix} 0.030 & -0.040 \\ 0.053 & 0.040 \end{bmatrix} \begin{bmatrix} 48.54 \\ -364.08 \end{bmatrix} = \begin{bmatrix} 16.0 \\ -12.0 \end{bmatrix}
$$

16.7 ANALYSIS OF FLEXURAL STRUCTURES BY THE DIRECT STIFFNESS METHOD

To begin, let us restrict ourselves to structures whose joints can rotate but do not translate. The continuous beam in Fig. 16.9a is an example of such a structure. Each joint is free to rotate, but none of the joints are able to translate. Since rotations are the only deformations that can take place at the ends of the individual members as well as at the structure joints, we can limit ourselves to the element and structure coordinates shown in Fig. 16.9b.

As a first step, let us construct a generally applicable element-stiffness matrix in structure coordinates. For a flexural element whose end rotations are δ_a and δ_b and whose end moments are q_a and q_b, as shown in Fig. 16.10, the element-stiffness

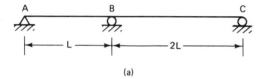

(a)

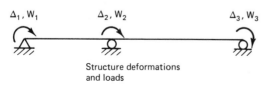

Structure deformations
and loads

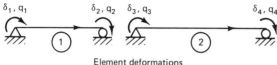

Element deformations
and loads

(b)

FIGURE 16.9

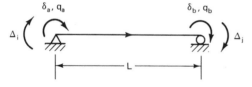

FIGURE 16.10

matrix in element coordinates is, according to Eq. (16.5), given by

$$\begin{bmatrix} q_a \\ q_b \end{bmatrix} = \frac{2EI}{L} \begin{bmatrix} 2 & 1 \\ 1 & 2 \end{bmatrix} \begin{bmatrix} \delta_a \\ \delta_b \end{bmatrix} \tag{16.30}$$

In a manner similar to that employed for trusses, we give the member a direction and denote the structure coordinates at the tail and head of the member by i and j, respectively. The transformation matrix $[T]_n$ for the element is then given by the equation

$$\begin{bmatrix} \delta_a \\ \delta_b \end{bmatrix} = \begin{bmatrix} 1 & 0 \\ 0 & 1 \end{bmatrix} \begin{bmatrix} \Delta_i \\ \Delta_j \end{bmatrix} \tag{16.31}$$

The above relation states that the element rotation at either end of the member is equal in magnitude and direction to the joint rotation at the same end.

Applying the relation $[K]_n = [T]_n^T [k]_n [T]_n$, which transforms the element-stiffness matrix from element to structure coordinates, to the matrices in Eqs. (16.30) and (16.31) leads to

$$[K]_n = \frac{2EI}{L} \begin{matrix} & i & j \\ \begin{bmatrix} 2 & 1 \\ 1 & 2 \end{bmatrix} & \begin{matrix} i \\ j \end{matrix} \end{matrix} \tag{16.32}$$

This is the element-stiffness matrix in structure coordinates for the flexural element in Fig. 16.10.

To illustrate the use of Eq. (16.32), let us form the structure-stiffness matrix for the beam in Fig. 16.9. The element-stiffness matrices for the two members are

$$[K]_1 = \frac{2EI}{L} \begin{matrix} & 1 & 2 \\ \begin{bmatrix} 2 & 1 \\ 1 & 2 \end{bmatrix} & \begin{matrix} 1 \\ 2 \end{matrix} \end{matrix}$$

and

$$[K]_2 = \frac{2EI}{L} \begin{matrix} & 2 & 3 \\ \begin{bmatrix} 1 & 0.5 \\ 0.5 & 1 \end{bmatrix} & \begin{matrix} 2 \\ 3 \end{matrix} \end{matrix}$$

Placing each term of these element-stiffness matrices into its appropriate position in the structure-stiffness matrix, as determined by the row and column designation of the term, we obtain

$$[K] = \frac{2EI}{L} \begin{matrix} & 1 & 2 & 3 \\ \begin{bmatrix} 2 & 1 & 0 \\ 1 & 3 & 0.5 \\ 0 & 0.5 & 1 \end{bmatrix} & \begin{matrix} 1 \\ 2 \\ 3 \end{matrix} \end{matrix}$$

To obtain a relation for calculating the member forces in a flexural structure, we start with Eq. (16.27),

$$[q]_n = [kT]_n [\Delta]_n$$

and substitute the matrices in Eqs. (16.30) and (16.31) for $[k]_n$ and $[T]_n$. This leads to

$$\begin{bmatrix} q_a \\ q_b \end{bmatrix} = \frac{2EI}{L} \begin{bmatrix} 2 & 1 \\ 1 & 2 \end{bmatrix} \begin{bmatrix} \Delta_i \\ \Delta_j \end{bmatrix} \tag{16.33}$$

Example 16.3

Determine the member forces for the structure in Fig. 16.11a.

We begin by numbering the members, choosing directions for them, and defining element and structure coordinates (Fig. 16.11b). Next we construct a table, as shown in Fig. 16.11c, that contains the data needed to form the $[K]_n$ matrices. Using these data and Eq. (16.32), we obtain

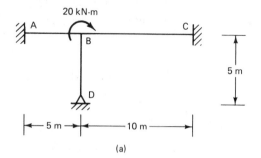

(a)

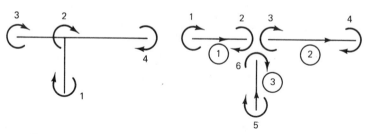

Structure coordinates   Element coordinates

(b)

Member	Length	i	j
1	5 m	3	2
2	10 m	2	4
3	5 m	1	2

(c)

FIGURE 16.11

$$[K]_1 = EI \begin{array}{cc} 3 & 2 \\ \begin{bmatrix} 0.8 & 0.4 \\ 0.4 & 0.8 \end{bmatrix} & \begin{array}{c} 3 \\ 2 \end{array} \end{array} \qquad [K]_2 = EI \begin{array}{cc} 2 & 4 \\ \begin{bmatrix} 0.4 & 0.2 \\ 0.2 & 0.4 \end{bmatrix} & \begin{array}{c} 2 \\ 4 \end{array} \end{array}$$

$$[K]_3 = EI \begin{array}{cc} 1 & 2 \\ \begin{bmatrix} 0.8 & 0.4 \\ 0.4 & 0.8 \end{bmatrix} & \begin{array}{c} 1 \\ 2 \end{array} \end{array}$$

Combining these matrices, we form the structure-stiffness matrix.

$$\begin{bmatrix} W_1 \\ W_2 \\ \hline W_3 \\ W_4 \end{bmatrix} = EI \begin{bmatrix} 0.8 & 0.4 & 0 & 0 \\ 0.4 & 2.0 & 0.4 & 0.2 \\ \hline 0 & 0.4 & 0.8 & 0 \\ 0 & 0.2 & 0 & 0.4 \end{bmatrix} \begin{bmatrix} \Delta_1 \\ \Delta_2 \\ \hline \Delta_3 \\ \Delta_4 \end{bmatrix}$$

Next we solve for the unknown structure deformations.

$$\begin{bmatrix} \Delta_1 \\ \Delta_2 \end{bmatrix} = \frac{1}{EI} \begin{bmatrix} 0.8 & 0.4 \\ 0.4 & 2.0 \end{bmatrix}^{-1} \begin{bmatrix} 0 \\ 20 \end{bmatrix} = \frac{1}{EI} \begin{bmatrix} -5.56 \\ 11.11 \end{bmatrix}$$

Finally, using Eq. (16.33), we determine the member forces.

$$\begin{bmatrix} q_1 \\ q_2 \\ q_3 \\ q_4 \\ q_5 \\ q_6 \end{bmatrix} = \begin{bmatrix} 0 & 0.4 & 0.8 & 0 \\ 0 & 0.8 & 0.4 & 0 \\ 0 & 0.4 & 0 & 0.2 \\ 0 & 0.2 & 0 & 0.4 \\ 0.8 & 0.4 & 0 & 0 \\ 0.4 & 0.8 & 0 & 0 \end{bmatrix} \begin{bmatrix} -5.56 \\ 11.11 \\ 0 \\ 0 \end{bmatrix} = \begin{bmatrix} 4.44 \\ 8.89 \\ 4.44 \\ 2.22 \\ 0 \\ 6.67 \end{bmatrix}$$

Loads Between Nodal Points

Up to now we have assumed all structure loads to be applied only at the joints or nodal points. Although this is satisfactory for trusses, it is obviously not the case for beams and frames. The latter are usually subjected to concentrated or distributed loads acting along the spans of the individual members. One way of dealing with this situation would be to subdivide each of the members on which intermediate loads are acting into a number of smaller elements and then approximate the given loading by a series of concentrated loads at the newly formed nodal points. However, this solution to the problem is very inefficient, and a more direct way of dealing with the situation will consequently be introduced.

The procedure that we shall employ for dealing with loads acting between nodal points is basically identical to the one we have already used in both the slope-deflection and moment-distribution methods. The element forces [q] are assumed to consist of two parts:

1. The element forces that are caused by the applied loads acting on a structure all of whose joints are assumed to be fixed.
2. The element forces due to the joint rotations and translations that actually take place.

For example, let us consider the continuous beam in Fig. 16.12a, which has a uniformly distributed load acting along one of its members. First, we artificially fix all the joints of the structure as shown in Fig. 16.12b. This necessitates the application of fictitious structure forces $[W_F]$. The element forces existing in this structure are what we used to refer to as fixed-end moments and shall now call $[q_F]$. Since the artificial forces $[W_F]$ do not exist in the actual structure, we now introduce a second system, shown in Fig. 16.12c, to which we apply structure loads $[W_E]$ that are equal and opposite to the forces $[W_F]$. We use the symbol $[q_E]$ to refer to the member forces produced by $[W_E]$.

Since the actual structure, in Fig. 16.12a, can be obtained by adding the structures in Figs. 16.12b and 16.12c, it follows that the member forces $[q]$ in the actual structure are given by the relation

$$[q] = [q_F] + [q_E] \tag{16.34}$$

The following example will serve to illustrate the above procedure.

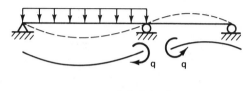

Actual structure

(a)

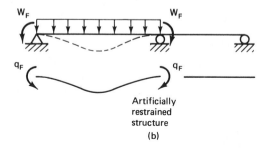

Artificially
restrained
structure

(b)

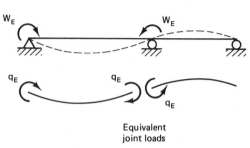

Equivalent
joint loads

(c)   **FIGURE 16.12**

Example 16.4

Determine the member forces for the structure in Fig. 16.13a.

As always, we begin by numbering the members, choosing directions for them, and defining element and structure coordinates (Fig. 16.13b).

When dealing with loads applied between nodal points, the solution consists of two parts. In the first we fix all joints and calculate the fixed-end member forces $[q_F]$ and the external loads $[W_F]$ needed to keep the joints from rotating. In the second part we determine the member forces $[q_E]$ due to the applied loads $[W_E]$.

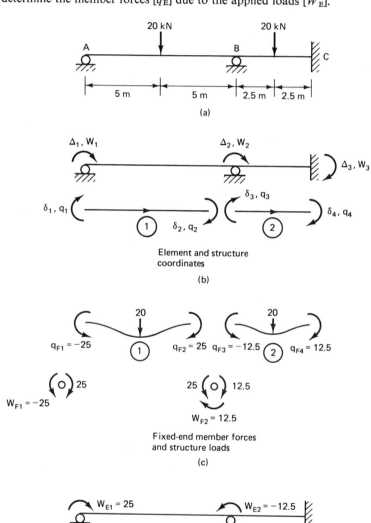

Element and structure
coordinates

(b)

Fixed-end member forces
and structure loads

(c)

Equivalent joint loads

(d)

FIGURE 16.13

(1) For the given applied loads the fixed-end member forces and structure loads are

$$[q_F] = \begin{bmatrix} -25 \\ 25 \\ -12.5 \\ 12.5 \end{bmatrix} \quad [W_F] = \begin{bmatrix} -25 \\ 12.5 \end{bmatrix}$$

(2) The structure loads $[W_E]$ are equal and opposite to $[W_F]$. Hence

$$[W_E] = \begin{bmatrix} 25 \\ -12.5 \end{bmatrix}$$

To calculate the member forces $[q_E]$ corresponding to these loads, we follow the procedure developed in the first part of this section.

The element-stiffness matrices are

$$[K]_1 = EI \begin{matrix} \;\;1\;\;\;\;\;2 \\ \begin{bmatrix} 0.4 & 0.2 \\ 0.2 & 0.4 \end{bmatrix} \begin{matrix} 1 \\ 2 \end{matrix} \end{matrix} \qquad [K]_2 = EI \begin{matrix} \;\;2\;\;\;\;\;3 \\ \begin{bmatrix} 0.8 & 0.4 \\ 0.4 & 0.8 \end{bmatrix} \begin{matrix} 2 \\ 3 \end{matrix} \end{matrix}$$

and the structure-stiffness matrix, obtained by combining the above, is given by

$$[K] = EI \begin{matrix} \;\;\;1\;\;\;\;\;\;\;2\;\;\;\;\;\;\;3 \\ \begin{bmatrix} 0.4 & 0.2 & 0 \\ 0.2 & 1.2 & 0.4 \\ 0 & 0.4 & 0.8 \end{bmatrix} \begin{matrix} 1 \\ 2 \\ 3 \end{matrix} \end{matrix}$$

Solving for the unknown joint displacements, we obtain

$$\begin{bmatrix} \Delta_1 \\ \Delta_2 \end{bmatrix} = \frac{1}{EI} \begin{bmatrix} 0.4 & 0.2 \\ 0.2 & 1.2 \end{bmatrix}^{-1} \begin{bmatrix} 25 \\ -12.5 \end{bmatrix} = \frac{1}{EI} \begin{bmatrix} 73.87 \\ -22.73 \end{bmatrix}$$

The member forces $[q_E]$ are calculated next.

$$\begin{bmatrix} q_{1E} \\ q_{2E} \\ q_{3E} \\ q_{4E} \end{bmatrix} = \begin{bmatrix} 0.4 & 0.2 & 0 \\ 0.2 & 0.4 & 0 \\ 0 & 0.8 & 0.4 \\ 0 & 0.4 & 0.8 \end{bmatrix} \begin{bmatrix} 73.87 \\ -22.73 \\ 0 \end{bmatrix} = \begin{bmatrix} 25.0 \\ 5.68 \\ -18.18 \\ -9.09 \end{bmatrix}$$

Finally we determine the member forces $[q]$ in the actual structure by adding $[q_F]$ and $[q_E]$. Thus

$$\begin{bmatrix} q_1 \\ q_2 \\ q_3 \\ q_4 \end{bmatrix} = \begin{bmatrix} -25.0 \\ 25.0 \\ -12.5 \\ +12.5 \end{bmatrix} + \begin{bmatrix} 25.0 \\ 5.68 \\ -18.18 \\ -9.09 \end{bmatrix} = \begin{bmatrix} 0 \\ 30.68 \\ -30.68 \\ 3.41 \end{bmatrix}$$

PROBLEMS

16.1. Using the method outlined in Section 16.4, form the $[K]$ matrix for the truss shown in the figure. Employ Eq. (16.11), $[K] = [T]^T[k_c][T]$, where $[T]$ is the transformation matrix for the entire structure and $[k_c]$ is the composite element-stiffness matrix.

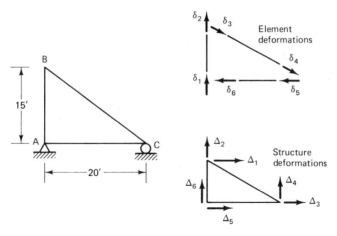

16.2. Form the $[K]$ matrix for the truss in Problem 16.1, using the procedure outlined in Section 16.5. A separate transformation matrix $[T]_n$ is formed for each member, and Eq. (16.16), $[K]_n = [T]_n^T[k]_n[T]_n$, is used to construct a $[K]_n$ matrix for each element. $[K]$ is obtained by combining the separate $[K]_n$ matrices.

16.3. Form the $[K]$ matrix for the truss in Problem 16.1, using the procedure outlined in Section 16.6 and illustrated in Example 16.1. Equation (16.22) is used to construct the element-stiffness matrices, which are combined to form $[K]$.

16.4 through 16.6. Use the direct stiffness method, outlined in Section 16.6 and illustrated in Example 16.2, to determine the unknown structure displacements and reactions, and the element forces.

16.4.

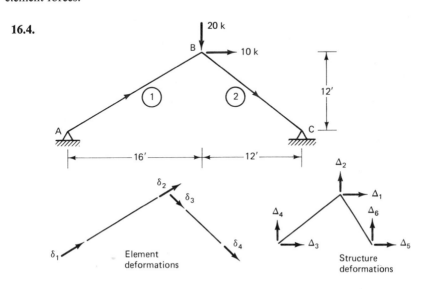

$A = \text{constant} = 1 \text{ in}^2$
$E = 30 \times 10^3 \text{ ksi}$

16.5.

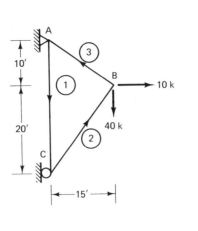

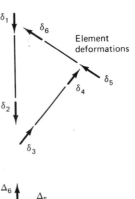

Element deformations

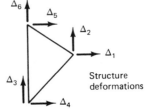

Structure deformations

$A = $ constant $ = 2$ in^2
$E = 30 \times 10^3$ ksi

16.6.

Element deformations

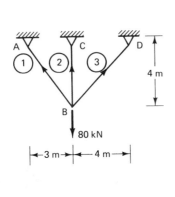

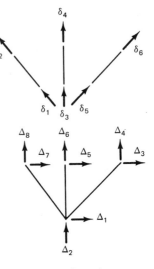

Structure deformations

$A = $ constant $ = 400$ mm^2
$E = 200 \times 10^6$ kN/m^2

16.7 and 16.8. Use the 2×2 element-stiffness matrix given by Eq. (16.32) to determine the element forces.

16.7.

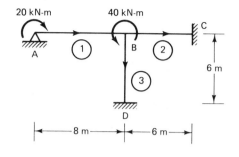

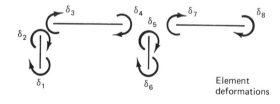

Element deformations

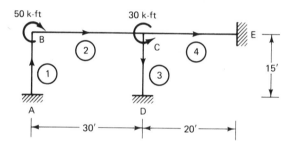

Structure deformations

$I = $ constant

16.8.

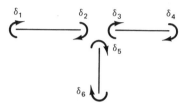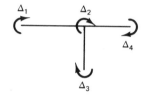

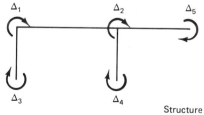

$I = $ constant

16.9 and 16.10. Use equivalent joint loads to determine the member forces.

16.9.

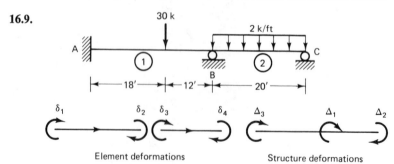

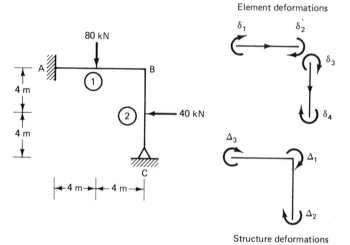

Element deformations   Structure deformations

$I = $ constant

16.10.

Element deformations

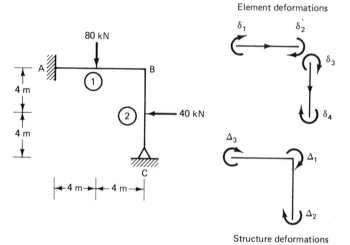

$I = $ constant

APPENDIX:

Matrix Algebra

A.1 DEFINITIONS

A *matrix* is a rectangular array of numbers, with m rows and n columns, for which certain arithmetic operations have been defined. It is customary to enclose the array within a set of brackets:

$$[A] = \begin{bmatrix} a_{11} & a_{12} & a_{13} \\ a_{21} & a_{22} & a_{23} \\ a_{31} & a_{32} & a_{33} \end{bmatrix}$$

The numbers $a_{11}, a_{12}, \ldots$ that make up the array are referred to as the *elements* of the matrix. The first subscript of any element denotes the row and the second subscript the column in which the element is located. Thus the element in the ith row and jth column is represented by a_{ij}. Either the capital letter used to name the matrix or the array itself can be used to represent the matrix.

The *order* or *size* of a matrix refers to the number of rows and columns that the matrix possesses. A matrix with m rows and n columns is said to be of order m by n.

A.2 TYPES OF MATRICES

Several special types of matrices are defined below.

Row matrix.   A *row matrix* consists of a single row.

$$[B] = [1 \quad 2 \quad 7 \quad 4]$$

Column matrix.   A *column matrix* consists of a single column.

$$[C] = \begin{bmatrix} 5 \\ 2 \\ -4 \end{bmatrix}$$

Square matrix.   A *square matrix* has the same number of rows as columns.

$$[D] = \begin{bmatrix} 4 & 2 & 1 \\ 6 & 0 & 2 \\ 2 & -1 & 12 \end{bmatrix}$$

Diagonal matrix.   A *diagonal matrix* is a square matrix that has zero elements everywhere except along the main diagonal. (The *main diagonal* runs from the upper left-hand corner to the lower right-hand corner of the matrix.)

$$[E] = \begin{bmatrix} 2 & 0 & 0 \\ 0 & 1 & 0 \\ 0 & 0 & 7 \end{bmatrix}$$

Symmetric matrix.   A *symmetric matrix* is a square matrix whose elements are symmetric about the main diagonal. Thus $a_{ji} = a_{ij}$ for a symmetric matrix.

$$[F] = \begin{bmatrix} 3 & 7 & -2 \\ 7 & 15 & 4 \\ -2 & 4 & 10 \end{bmatrix}$$

Null matrix.   The *null matrix* is a matrix all of whose elements are zero. The null matrix can be likened to the numeral zero in ordinary algebra.

$$[G] = \begin{bmatrix} 0 & 0 & 0 \\ 0 & 0 & 0 \\ 0 & 0 & 0 \end{bmatrix}$$

Identity matrix.   The *identity matrix* is a diagonal matrix all of whose elements are ones. It is usually represented by [I], and it serves the same purpose in matrix algebra that the number one does in ordinary algebra.

$$[I] = \begin{bmatrix} 1 & 0 & 0 \\ 0 & 1 & 0 \\ 0 & 0 & 1 \end{bmatrix}$$

A.3 EQUALITY, ADDITION, AND SUBTRACTION

Equality. For two matrices to be equal they must be identical term by term. Thus two equal matrices will necessarily be of the same order.

$$[A] = \begin{bmatrix} 3 & 6 \\ 4 & -2 \\ 1 & 7 \end{bmatrix}, \qquad [B] = \begin{bmatrix} 3 & 6 \\ 4 & -2 \\ 1 & 7 \end{bmatrix}$$

$$[A] = [B]$$

Addition and subtraction. Addition or subtraction of matrices is possible only if the matrices are of the same order. The addition or subtraction is carried out by adding or subtracting corresponding elements term by term.

$$\begin{bmatrix} 3 & 1 & 4 \\ 6 & -7 & 2 \end{bmatrix} + \begin{bmatrix} 2 & 0 & -1 \\ 6 & 2 & 1 \end{bmatrix} = \begin{bmatrix} 5 & 1 & 3 \\ 12 & -5 & 3 \end{bmatrix}$$

It should be noted that the matrix obtained by adding or subtracting two matrices is of the same order as the original matrices.

A.4 MULTIPLICATION

Multiplication of a scalar by a matrix. The product of a scalar and a matrix is obtained by multiplying each element of the matrix by the scalar.

$$3 \begin{bmatrix} 2 & 4 \\ 6 & 2 \end{bmatrix} = \begin{bmatrix} 6 & 12 \\ 18 & 6 \end{bmatrix}$$

Multiplication of two matrices. The rule governing the multiplication of two matrices has been chosen in such a way that certain operations, such as the solution of simultaneous equations, that occur frequently in ordinary algebra can be easily carried out in matrix algebra. Thus, let us consider the following two simultaneous equations:

$$a_{11}x_1 + a_{12}x_2 = c_1$$
$$a_{21}x_1 + a_{22}x_2 = c_2 \tag{A.1}$$

Here x_1 and x_2 are the unknowns and the a's and c's are constants. Next we define the three matrices

$$[A] = \begin{bmatrix} a_{11} & a_{12} \\ a_{21} & a_{22} \end{bmatrix} \qquad [X] = \begin{bmatrix} x_1 \\ x_2 \end{bmatrix} \qquad [C] = \begin{bmatrix} c_1 \\ c_2 \end{bmatrix}$$

and write

$$\begin{bmatrix} a_{11} & a_{12} \\ a_{21} & a_{22} \end{bmatrix} \begin{bmatrix} x_1 \\ x_2 \end{bmatrix} = \begin{bmatrix} c_1 \\ c_2 \end{bmatrix} \tag{A.2}$$

Our objective is to define matrix multiplication in such a way that the algebraic equations (A.1) can be replaced by the matrix equation (A.2). This will be possible if c_1, the element in *row one and column one* of $[C]$, is obtained by multiplying the *first row in* $[A]$ by the *first column in* $[X]$ term by term, and adding the resulting products. That is, $c_1 = a_{11}x_1 + a_{12}x_2$. In a similar manner, c_2, the element occupying the *second row and first column* in $[C]$ is obtained by multiplying the *second row in* $[A]$ by the *first column in* $[X]$ term by term, and adding the resulting products. Thus $c_2 = a_{21}x_1 + a_{22}x_2$.

In general, if

$$[A][B] = [C]$$

then any element c_{ij} in $[C]$ is obtained by multiplying, term by term, the ith row in $[A]$ by the jth column in $[B]$ and adding the resulting products. This rule can be expressed analytically as

$$c_{ij} = \sum_{k=1}^{n} a_{ik}b_{kj} \tag{A.3}$$

where n is the number of columns in $[A]$ and the number of rows in $[B]$.

For example,

$$\begin{bmatrix} 1 & 3 & 2 \\ 0 & -2 & 5 \\ 4 & 1 & -3 \end{bmatrix} \begin{bmatrix} 3 & 1 \\ 0 & 4 \\ -1 & 2 \end{bmatrix} = \begin{bmatrix} 1 & 17 \\ -5 & 2 \\ 15 & 2 \end{bmatrix}$$

The rule for multiplication given by Eq. (A.3) immediately establishes the fact that two matrices can be multiplied only if the number of columns in the first is equal to the number of rows in the second. Two matrices that satisfy this criterion are said to be *conformable for multiplication*. There are no restrictions regarding the number of rows in the first matrix and the number of columns in the second. The product of two matrices will have the same number of rows as the first matrix and the same number of columns as the second.

A simple way of checking the conformability of two matrices and to determine the size of their product is to write the order of the matrices below them, as follows:

$$\begin{array}{ccc} [A] & [B] & = & [C] \\ 4 \times 3 & 3 \times 2 & & 4 \times 2 \end{array}$$

Matrix multiplication is associative and distributive. Hence

$$[A][BC] = [AB][C]$$

and

$$[A][B + C] = [AB] + [AC]$$

However, matrix multiplication is not generally commutative. Thus

$$[A][B] \neq [B][A]$$

An exception to the last rule occurs when one of the two matrices to be multiplied is the identity matrix. For example,

$$\begin{bmatrix} 2 & 3 \\ 2 & 4 \end{bmatrix}\begin{bmatrix} 1 & 0 \\ 0 & 1 \end{bmatrix} = \begin{bmatrix} 2 & 3 \\ 2 & 4 \end{bmatrix}$$

$$\begin{bmatrix} 1 & 0 \\ 0 & 1 \end{bmatrix}\begin{bmatrix} 2 & 3 \\ 2 & 4 \end{bmatrix} = \begin{bmatrix} 2 & 3 \\ 2 & 4 \end{bmatrix}$$

This example also illustrates the fact that multiplying by the identity matrix in matrix algebra is analogous to multiplying by one in ordinary algebra.

A.5 TRANSPOSE OF A MATRIX

The *transpose* of a matrix is obtained by interchanging the rows and columns of the matrix. For example, if

$$[A] = \begin{bmatrix} 3 & 2 & -1 \\ 6 & 2 & 4 \end{bmatrix}$$

Then $[A]^{\mathrm{T}}$, the transpose of $[A]$, is

$$[A]^{\mathrm{T}} = \begin{bmatrix} 3 & 6 \\ 2 & 2 \\ -1 & 4 \end{bmatrix}$$

The operation of transposing a matrix has no equivalent in ordinary algebra. It arises in matrix algebra because matrix operations can only be performed if the matrices being operated on are of the proper order.

To see how the operation of transposing a matrix makes two matrices conformable, let us consider the work W performed by a force F moving through a distance D. In order to calculate W, we define two matrices $[F]$ and $[D]$ consisting of the x, y, and z components of the force and the distance. Thus

$$[F] = \begin{bmatrix} f_x \\ f_y \\ f_z \end{bmatrix} \qquad [D] = \begin{bmatrix} d_x \\ d_y \\ d_z \end{bmatrix}$$

Although it seems natural that the work should be the product of $[F]$ and $[D]$, these matrices as defined above are not conformable and thus cannot be multiplied. However, if we form the transpose of $[F]$, it can be multiplied by $[D]$ and the product is indeed equal to the work. Thus

$$[W] = [F]^T[D] = [f_x \quad f_y \quad f_z]\begin{bmatrix} d_x \\ d_y \\ d_z \end{bmatrix} = [f_x d_x + f_y d_y + f_z d_z]$$

A.6 DETERMINANTS

Since determinants are involved in the inversion of matrices, which we will consider next, let us note a few fundamentals about determinants. Determinants are closely related to the solution of simultaneous equations. For example, consider the following two equations, in which x_1 and x_2 are the unknowns:

$$a_{11}x_1 + a_{12}x_2 = c_1$$
$$a_{21}x_1 + a_{22}x_2 = c_2$$

To solve for x_1, we multiply the first equation by a_{22} and the second equation by $-a_{12}$ and add the resulting equations. Thus

$$a_{11}a_{22}x_1 + a_{12}a_{22}x_2 = c_1 a_{22}$$
$$\underline{-a_{21}a_{12}x_1 - a_{22}a_{12}x_2 = -c_2 a_{12}}$$
$$x_1(a_{11}a_{22} - a_{21}a_{12}) = c_1 a_{22} - c_2 a_{12}$$

from which

$$x_1 = \frac{c_1 a_{22} - c_2 a_{12}}{a_{11}a_{22} - a_{21}a_{12}} \tag{A.4}$$

The result given by Eq. (A.4) can also be expressed as the quotient of two determinants. That is,

$$x_1 = \frac{\begin{vmatrix} c_1 & a_{12} \\ c_2 & a_{22} \end{vmatrix}}{\begin{vmatrix} a_{11} & a_{12} \\ a_{21} & a_{22} \end{vmatrix}} \tag{A.5}$$

The first fact to note about determinants is that they consist of square arrays of numbers bounded by two vertical lines. Next we define the value of a determinant in such a way that the solution given by Eq. (A.5) is identical to that given by Eq. (A.4). This can be accomplished if the value of a 2×2 determinant is equal to the product of the elements along its major diagonal minus the product of the elements along the minor diagonal. Finally, we note that the determinant in the denominator of Eq. (A.5) consists of the array of coefficients a_{ij} of the original simultaneous equations and that the determinant in the numerator contains the same array except that the coefficients of the unknown x_1 have been replaced by the constants c_i.

In working with determinants and matrices it is essential to keep in mind the basic differences between the two. A determinant is always square and can be reduced to a single value, whereas a matrix is simply an array of numbers, not necessarily square, which cannot be reduced to a single value.

The form of solution given by Eq. (A.5) is known as *Cramer's rule* and is applicable to any number of simultaneous equations. However, the method of evaluating a determinant described above can be applied only to a 2×2 determinant. Therefore let us now consider a more general procedure for evaluating determinants, which is applicable to determinants of any size. The method is called the *Laplace expansion* of determinants, and it makes use of the concepts of minors and cofactors.

Minors.   The *minor* of an element in a determinant is the determinant that remains when the row and column containing the element are removed. For example, suppose we are given the determinant

$$|A| = \begin{vmatrix} a_{11} & a_{12} & a_{13} \\ a_{21} & a_{22} & a_{23} \\ a_{31} & a_{32} & a_{33} \end{vmatrix} \tag{A.6}$$

The minor of element a_{11} is

$$M_{11} = \begin{vmatrix} a_{22} & a_{23} \\ a_{32} & a_{33} \end{vmatrix}$$

Similarly, the minor of element a_{12} is

$$M_{12} = \begin{vmatrix} a_{21} & a_{23} \\ a_{31} & a_{33} \end{vmatrix}$$

Cofactors.   The *cofactor* of an element in a determinant is the minor of the element multiplied by $-1^{(i+j)}$, where i and j are the row and column designation of the element. Thus, the minor is multiplied by $+1$ if the sum of the row and column designation of the elements is even and by -1 if the sum is odd. According to this definition, the cofactor of a_{11} in the determinant given by (A.6) is

$$C_{11} = (-1)^2 M_{11} = + \begin{vmatrix} a_{22} & a_{23} \\ a_{32} & a_{33} \end{vmatrix}$$

and the cofactor of a_{12} is

$$C_{12} = (-1)^3 M_{12} = - \begin{vmatrix} a_{21} & a_{23} \\ a_{31} & a_{33} \end{vmatrix}$$

Laplace expansion of determinants.   The value of a determinant is equal to the sum of the products of the elements and their cofactors in any row or column of the determinant. Thus the value of an $n \times n$ determinant expanded along one of its rows is given by

$$|A| = \sum_{j=1}^{n} a_{ij} C_{ij} \quad \text{for any } i$$

As an illustration let us expand the following determinant along its first row:

$$|A| = \begin{vmatrix} 1 & 2 & -4 \\ 3 & 1 & -2 \\ 4 & 2 & 3 \end{vmatrix}$$

We obtain

$$|A| = (1) \begin{vmatrix} 1 & -2 \\ 2 & 3 \end{vmatrix} - (2) \begin{vmatrix} 3 & -2 \\ 4 & 3 \end{vmatrix} + (-4) \begin{vmatrix} 3 & 1 \\ 4 & 2 \end{vmatrix}$$

$$= (1)(7) - (2)(17) + (-4)(2) = -35$$

A.7 INVERSE OF A MATRIX

Matrix algebra does not include the operation of division. Instead of dividing a matrix [A] by a matrix [B], one multiplies [A] by the inverse of [B]. The *inverse* of a matrix $[B]$ is defined as the matrix which when multiplied by $[B]$ produces the identity matrix. That is,

$$[B][B]^{-1} = [B]^{-1}[B] = [I]$$

where $[B]^{-1}$ denotes the inverse of $[B]$. The inverse of a matrix can be likened to the reciprocal of a number in ordinary algebra. Only a nonsingular square matrix has an inverse.

Since the operation of inverting a matrix is closely related to the solution of simultaneous equations, let us consider the following set of equations:

$$a_{11}x_1 + a_{12}x_2 + a_{13}x_3 = c_1$$
$$a_{21}x_1 + a_{22}x_2 + a_{23}x_3 = c_2 \qquad \text{(A.7)}$$
$$a_{31}x_1 + a_{32}x_2 + a_{33}x_3 = c_3$$

The equations can be written in matrix form as

$$\begin{bmatrix} a_{11} & a_{12} & a_{13} \\ a_{21} & a_{22} & a_{23} \\ a_{31} & a_{32} & a_{33} \end{bmatrix} \begin{bmatrix} x_1 \\ x_2 \\ x_3 \end{bmatrix} = \begin{bmatrix} c_1 \\ c_2 \\ c_3 \end{bmatrix} \qquad \text{(A.8)}$$

or simply as

$$[A][X] = [C] \qquad \text{(A.9)}$$

where $[A]$ is the matrix of coefficients, $[X]$ the matrix containing the unknowns, and $[C]$ the matrix made up of the constants on the right-hand sides of the equations.

To solve for $[X]$, we premultiply both sides of Eq. (A.9) by $[A]^{-1}$. Thus

$$[A]^{-1}[A][X] = [A]^{-1}[C]$$

from which

$$[X] = [A]^{-1}[C] \qquad \text{(A.10)}$$

It is thus evident that one can obtain the solution of a set of simultaneous equations if one can determine the inverse of the coefficient matrix $[A]$. To see how the inverse of $[A]$ is obtained, let us solve the simultaneous equations in (A.7) using Cramer's rule.

$$x_1 = \frac{\begin{vmatrix} c_1 & a_{12} & a_{13} \\ c_2 & a_{22} & a_{23} \\ c_3 & a_{32} & a_{33} \end{vmatrix}}{|A|} \qquad x_2 = \frac{\begin{vmatrix} a_{11} & c_1 & a_{13} \\ a_{21} & c_2 & a_{23} \\ a_{31} & c_3 & a_{33} \end{vmatrix}}{|A|}$$

$$x_3 = \frac{\begin{vmatrix} a_{11} & a_{12} & c_1 \\ a_{21} & a_{22} & c_2 \\ a_{31} & a_{32} & c_3 \end{vmatrix}}{|A|}$$

where $|A|$ is the determinant of the coefficients a_{ij}. Expansion of the numerator of x_1 along the first column, the numerator of x_2 along the second column, and the numerator of x_3 along the third column gives

$$x_1 = \frac{1}{|A|}(c_1C_{11} + c_2C_{21} + c_3C_{31})$$

$$x_2 = \frac{1}{|A|}(c_1C_{12} + c_2C_{22} + c_3C_{32}) \qquad \text{(A.11)}$$

$$x_3 = \frac{1}{|A|}(c_1C_{13} + c_2C_{23} + c_3C_{33})$$

where C_{ij} is the cofactor of the element in the ith row and jth column. In the problem we are considering, C_{11} happens to be not only the cofactor of c_1 in the numerator of x_1 but also the cofactor of a_{11} in $|A|$. Thus C_{ij} is in general equal to the cofactor of a_{ij} in $|A|$.

Rewriting (A.11) in matrix from gives

$$\begin{bmatrix} x_1 \\ x_2 \\ x_3 \end{bmatrix} = \frac{1}{|A|} \begin{bmatrix} C_{11} & C_{21} & C_{31} \\ C_{12} & C_{22} & C_{32} \\ C_{13} & C_{23} & C_{33} \end{bmatrix} \begin{bmatrix} c_1 \\ c_2 \\ c_3 \end{bmatrix} \qquad \text{(A.12)}$$

The square matrix on the right-hand side of Eq. (A.12) is called the *adjoint matrix*. It can be obtained in the following way. First we form the matrix $[A]$ containing the coefficients of the simultaneous equations.

$$[A] = \begin{bmatrix} a_{11} & a_{12} & a_{13} \\ a_{21} & a_{22} & a_{23} \\ a_{31} & a_{32} & a_{33} \end{bmatrix}$$

Next we replace every element in $[A]$ by its cofactor. This new matrix is referred to as the *cofactor matrix*.

$$[\text{Cof. A}] = \begin{bmatrix} C_{11} & C_{12} & C_{13} \\ C_{21} & C_{22} & C_{23} \\ C_{31} & C_{32} & C_{33} \end{bmatrix}$$

Finally, we form the adjoint matrix by taking the transpose of the cofactor matrix.

$$[\text{Adj. } A] = \begin{bmatrix} C_{11} & C_{21} & C_{31} \\ C_{12} & C_{22} & C_{32} \\ C_{13} & C_{23} & C_{33} \end{bmatrix}$$

Comparison of Eqs. (A.12) and (A.10) indicates that the inverse of a matrix is equal to its adjoint matrix divided by its determinant.

To illustrate the foregoing ideas, let us solve the following set of simultaneous equations, using the inverse of the coefficient matrix:

$$x_1 + x_2 + x_3 = 2$$
$$2x_1 - x_2 + x_3 = 5$$
$$x_1 + 2x_2 + 2x_3 = 3$$

$$[A] = \begin{bmatrix} 1 & 1 & 1 \\ 2 & -1 & 1 \\ 1 & 2 & 2 \end{bmatrix}$$

$$[\text{Cof. } A] = \begin{bmatrix} -4 & -3 & 5 \\ 0 & 1 & -1 \\ 2 & 1 & -3 \end{bmatrix}$$

$$[\text{Adj. } A] = \begin{bmatrix} -4 & 0 & 2 \\ -3 & 1 & 1 \\ 5 & -1 & -3 \end{bmatrix}$$

$$|A| = 1(-4) + 1(-3) + 1(5) = -2$$

$$\begin{bmatrix} x_1 \\ x_2 \\ x_3 \end{bmatrix} = \frac{1}{-2} \begin{bmatrix} -4 & 0 & 2 \\ -3 & 1 & 1 \\ 5 & -1 & -3 \end{bmatrix} \begin{bmatrix} 2 \\ 5 \\ 3 \end{bmatrix} = \begin{bmatrix} 1 \\ -1 \\ 2 \end{bmatrix}$$

A.8 PARTITIONING OF MATRICES

Sometimes it is desirable to subdivide a matrix into several parts called *submatrices*. This process, known as *partitioning*, is accomplished by using horizontal and vertical lines as follows

$$[A] = \begin{bmatrix} 3 & 2 & \vdots & 6 & 2 \\ 1 & 4 & \vdots & 1 & 1 \\ \cdots & \cdots & & \cdots & \cdots \\ 2 & 7 & \vdots & 4 & 1 \\ 3 & 2 & \vdots & 1 & 7 \end{bmatrix} = \begin{bmatrix} A_{11} & A_{12} \\ A_{21} & A_{22} \end{bmatrix}$$

where A_{11}, A_{12}, A_{21}, and A_{22} are submatrices of $[A]$.

To illustrate how one carries out the basic operations of matrix algebra on partitioned matrices, let us partition two matrices, $[A]$ and $[B]$, and obtain their product $[C]$.

$$[A] = \begin{bmatrix} 3 & 2 & \vdots & 1 \\ 1 & 4 & \vdots & 2 \\ \cdots & \cdots & & \cdots \\ 4 & 2 & \vdots & 1 \end{bmatrix} = \begin{bmatrix} A_{11} & A_{12} \\ A_{21} & A_{22} \end{bmatrix}$$

$$[B] = \begin{bmatrix} 4 & 2 \\ 3 & 6 \\ \cdots & \cdots \\ 1 & 2 \end{bmatrix} = \begin{bmatrix} B_{11} \\ B_{21} \end{bmatrix}$$

Initially we multiply [A] and [B], treating the submatrices as elements.

$$[A][B] = \begin{bmatrix} A_{11} & A_{12} \\ A_{21} & A_{22} \end{bmatrix}\begin{bmatrix} B_{11} \\ B_{21} \end{bmatrix} = \begin{bmatrix} C_{11} \\ C_{21} \end{bmatrix} \tag{A.13}$$

where $C_{11} = A_{11}B_{11} + A_{12}B_{21}$ and $C_{21} = A_{21}B_{11} + A_{22}B_{21}$.

Next we calculate C_{11} and C_{21} by operating on the submatrices. Thus

$$A_{11}B_{11} = \begin{bmatrix} 3 & 2 \\ 1 & 4 \end{bmatrix}\begin{bmatrix} 4 & 2 \\ 3 & 6 \end{bmatrix} = \begin{bmatrix} 18 & 18 \\ 16 & 26 \end{bmatrix}$$

$$A_{12}B_{21} = \begin{bmatrix} 1 \\ 2 \end{bmatrix}[1 \quad 2] = \begin{bmatrix} 1 & 2 \\ 2 & 4 \end{bmatrix}$$

$$C_{11} = A_{11}B_{11} + A_{12}B_{21} = \begin{bmatrix} 19 & 20 \\ 18 & 30 \end{bmatrix}$$

$$A_{21}B_{11} = [4 \quad 2]\begin{bmatrix} 4 & 2 \\ 3 & 6 \end{bmatrix} = [22 \quad 20]$$

$$A_{22}B_{21} = [1][1 \quad 2] = [1 \quad 2]$$

$$C_{21} = A_{21}B_{11} + A_{22}B_{21} = [23 \quad 22]$$

Finally, we combine C_{11} and C_{21} to obtain [C], the product of [A] and [B].

$$[C] = \begin{bmatrix} 19 & 20 \\ 18 & 30 \\ \hline 23 & 22 \end{bmatrix}$$

To carry out an algebraic operation involving partitioned matrices requires that two conditions be satisfied. First, the matrices to be partitioned must be conformable, and second, the partitioning must be done in such a way that the operations involving the submatrices can be carried out. For example, if two matrices are to be added or subtracted, the vertical and horizontal partitioning of both must be identical. If two matrices are to be multiplied, the vertical partitioning of the first must be similar to the horizontal partioning of the second. Thus if the vertical partitioning of the first matrix divides it into two parts having three and two columns, respectively, then the horizontal partitioning of the second matrix must divide it into two parts having three and two rows, respectively. For multiplication the horizontal partitioning of the first matrix and the vertical partitioning of the second matrix are independent of one another.

Selected Answers
to Even-Numbered Problems

Chapter 2

2.2. $A_x = 30$ kN ←; $A_y = 65$ kN ↑; $B_y = 95$ kN ↑

2.4. $A_x = 2.28$ k ←; $A_y = 6.96$ k ↑; $B = 3.8$ k ↗

2.6. $A_y = 52.2$ k ↑; $B_x = 18$ →; $B_y = 16.8$ k ↑

2.8. $A_x = 40$ k ←; $A_y = 6$ k ↑; $B_y = 44$ k ↑

2.10. $A_y = 75$ kN ↑; $B_x = 0$; $B_y = 85$ kN ↑; $C_x = 0$; $C_y = 100$ kN ↑

2.12. $A_x = 10.4$ k ←; $A_y = 2.16$ k ↑; $B = 2.7$ k ↙

2.14. $A_y = 101.2$ kN ↑; $A_x = 31.8$ kN ←; $B_y = 58.8$ kN ↑; $B_x = 88.2$ kN ←

2.16. $A_y = 30$ kN ↑; $A_x = 30$ kN →; $B_y = 110$ kN ↑; $C_x = 30$ kN ←; $C_y = 20$ kN ↑

2.18. $A_y = 29.2$ k ↑; $A_x = 0$; $B_y = 86$ k ↑; $C_y = 54.8$ k ↑; $D_y = 15$ k ↑

Chapter 3

3.2. $AB = 12.5$ k (C); $AE = 20$ k (T); $BE = 0$; $BC = 12.5$ k (T); $CD = 7.5$ k (C); $BD = 25$ k (C); $DE = 20$ k (T)

3.4. $AB = 8$ k (C); $AE = 16$ k (T); $BE = 0$; $BC = 17.9$ k (C); $CE = 16$ k (C); $CD = 16$ k (C); $DE = 22.6$ k (T)

3.6. $AB = 43.3$ kN (T); $AF = 10$ kN (T); $BF = 16.7$ kN (C); $EF = 60$ kN (C); $BE = 45$ kN (C); $CE = 60$ kN (C); $BC = 54.1$ kN (T); $DE = 45$ kN (C); $CD = 54.1$ kN (T)

3.8. $BC = 42.7$ k (C); $BG = 13.3$ k (T); $HG = 32$ k (T)

3.10. $BC = 26.7$ k (C); $GC = 0$; $GF = 26.7$ k (T)

3.12. $BC = 37.3$ k (C); $BG = 2.4$ k (C); $HG = 35$ k (T)

3.14. $AB = 81.6$ kN (T); $BC = 15.5$ kN (C); $CD = 109$ kN (C); $DE = 81.6$ kN (C); $EF = 70.0$ kN (T); $FG = 73.1$ kN (T); $GH = 136$ kN (T); $DF = 21.0$ kN (C); $DG = 24.3$ kN (T); $CG = 64.0$ kN (T); $BG = 38.3$ kN (C); $BH = 39.0$ kN (C); $AH = 130$ kN (T)

3.16. $AB = 3$ k (T); $BC = 1.7$ kN (T); $CD = 2.0$ kN (T); $DE = 15$ kN (C); $EF = 0$; $AF = 8.0$ kN (T); $BF = 2.6$ kN (C); $CF = 14.3$ kN (C); $DF = 17.7$ kN (T)

3.18. $AB = 133.3$ kN (C); $BF = 28.6$ kN (C); $AF = 67.1$ kN (T); $CF = 72.8$ kN (T); $BC = 101$ kN (C); $CD = 86.8$ kN (C); $DF = 28.6$ kN (C); $DE = 90.9$ kN (C); $EF = 67.1$ kN (T)

Chapter 4

4.2. (Bar forces only) $AB = 83.4$ kN (T); $AC = 228$ kN (C); $AD = 172$ kN (T);
$BC = 151$ kN (T); $BD = 387$ kN (C); $CD = 89.0$ kN (T)
4.4. (Bar forces only) $AB = 27.3$ kN (C); $AD = 180$ kN (C); $AE = 419$ kN (C);
$BC = 43.6$ kN (T); $BE = 101$ kN (T); $CD = 27.1$ kN (C); $CE = 141$ kN (T); $DE = 141$ kN (T)

Chapter 5

5.2. $V_{max} = 12$ k; $M_{max} = 120$ k-ft   **5.4.** $V_{max} = 40$ kN; $M_{max} = -160$ kN-m
5.6. $V_{max} = 32$ k; $M_{max} = 171$ k-ft   **5.8.** $V_{max} = 11$ k; $M_{max} = -123$ k-ft
5.10. $V_{max} = -42.5$ kN; $M_{max} = -120$ kN-m   **5.12.** $V_{max} = -94.6$ kN; $M_{max} = 183$ kN-m
5.14. $V_{max} = -80$ k; $M_{max} = 616$ k-ft   **5.16.** $V_{max} = -6.13$ k; $M_{max} = 70.4$ k-ft
5.18. $V_{max} = 85$ kN; $M_{max} = 141$ kN-m   **5.20.** $V_{max} = -131$ kN; $M_{max} = 853$ kN-m
5.22. $V_{max} = 106$ k; $M_{max} = 600$ k-ft   **5.24.** $V_{max} = 80$ kN; $M_{max} = 267$ kN-m
5.26. $V_{max} = -260$ k; $M_{max} = 10,500$ k-ft

Chapter 6

6.2. $\dfrac{wx}{24EI}(L^3 - 2Lx^2 + x^3)$ **6.4.** $\dfrac{w_o x}{360LEI}(7L^4 - 10L^2x^2 + 3x^4)$

6.8. $\delta_B = \dfrac{wL^4}{8EI}$; $\theta_B = \dfrac{wL^3}{6EI}$ **6.10.** $\delta_C = 4.0$ in; $\theta_C = 0.0347$ rad

6.12. $\delta_B = 51.8$ mm; $\theta_A = 0.018$ rad   **6.14.** $\delta_A = 4.42$ in; $\delta_C = 1.66$ in   **6.16.** $\delta_B = 16.7$ mm

6.18. $\delta_D = 1.73$ in   **6.20.** $\delta_C = \dfrac{3PL^3}{32EI}$ **6.22.** $\delta_B = 1.95$ in   **6.24.** $\delta_B = 56.9$ mm

6.26. $\delta_B = 24$ mm; $\theta_A = 0.01$ rad   **6.28.** $\delta_C = 8.81$ in   **6.30.** $\delta_C = 23.3$ mm

6.32. $\delta_C = \dfrac{PL^3}{8EI}$ **6.34.** $\delta_A = 2.67$ mm; $\delta_C = 1.33$ mm

Chapter 7

7.2. $\delta_B = \dfrac{4PL^3}{243EI}$ **7.4.** $\delta_B = \dfrac{7PL^3}{486EI}$ **7.6.** $\delta_B = 2.5$ in   **7.8.** $\delta_D = 67.5$ mm

7.10. $\delta_E = 3.33$ in   **7.12.** $\delta_{AV} = 2.1$ in   **7.14.** $\delta_{CH} = 3.2$ in; $\theta_C = .013$ rad
7.16. $\theta_B = .044$ rad   **7.18.** $\delta_{EH} = 1.58$ in   **7.20.** $\delta_D = 0.55$ in   **7.22.** $\delta_C = 49$ mm
7.24. $\delta_C = 8.8$ mm

Chapter 9

9.2. $R_{AY} = 39.0$ k ↑ ; $R_{AX} = 32.5$ k → $R_{EY} = 21.0$ k ↑ ; $R_{EX} = 32.5$ k ←
9.4. $R_{AY} = 32.5$ k ↑ ; $R_{AX} = 35$ k → $R_{GY} = 37.5$ ↑ ; $R_{GX} = 35$ k ←
9.6. $R_{AY} = 4.6$ k ↑ ; $R_{AX} = 8.9$ k ← $R_{DY} = 7.4$ k ↑ ; $R_{DX} = 8.9$ k →

Chapter 11

11.2. $R_A = 6.2$ kN ↑ ; $M_A = 5$ kN-m; $R_B = 23.8$ kN ↑
11.4. $R_A = 40.5$ kN ↑ ; $R_B = 105.3$ kN ↑ ; $R_C = 14.2$ kN ↑
11.6. $R_A = 65.8$ kN ↑ ; $R_B = 129.6$ kN ↑ ; $R_C = 4.6$ kN ↑
11.8. $R_{AH} = 6.7$ kN → ; $R_{AV} = 31.1$ kN ↑ ; $R_{CH} = 6.7$ kN ← ; $R_{CV} = 8.9$ kN ↑
11.10. $R_A = 12$ kN ↓ ; $M_A = 8$ kN-m; $T_B = 32$ kN
11.12. $R_A = 16.4$ k ↑ ; $M_A = 127$ k-ft; $R_C = 3.6$ k ↑
11.14. $R_A = 0.075\ P$ ↓ ; $R_B = 0.575\ P$ ↑
11.16. $R_A = 52$ kN ↑ ; $R_B = 106$ kN ↑ ; $R_C = 200$ kN ↑ ; $R_D = 82$ kN ↑
11.18. $R_A = 1.3$ kN ↑ ; $R_B = 7.9$ kN ↓ ; $R_C = 56.6$ kN ↑
11.20. $AB = 30$ kN (T); $BC = 60$ kN (T); $CD = 30$ kN (C); $AD = 60$ kN (C);
$BD = 67.1$ kN (T); $AC = 67.1$ kN (C)

Chapter 12

12.2. $R_A = 75.7$ kN ↑ ; $M_A = 127.5$ kN-m; $R_B = 134.3$ kN ↑
12.4. $R_A = 10.8$ k ↑ ; $R_B = 35.9$ k ↑ ; $R_C = 8.3$ k ↑
12.6. $R_{AV} = 17.5$ kN ↑ ; $R_{AH} = 4$ kN → ; $R_{CV} = 12.5$ kN ↑ ; $R_{CH} = 4$ kN ←
12.8. $R_A = 92.9$ kN ↑ ; $R_B = 228.6$ kN ↑ ; $R_C = 78.6$ kN ↑ ; $M_A = 142.9$ kN-m
12.10. $R_{AV} = 53.1$ kN ↑ ; $R_{AH} = 35.8$ kN → ; $M_A = 105.2$ kN-m; $T_{BC} = 44.76$ kN
12.12. $R_{AV} = 5$ k ↑ ; $R_{AH} = 3.18$ k → ; $R_{CV} = 5$ k ↑ ; $R_{CH} = 3.18$ k ←
12.14. $M_{\text{MAX}} = 1000$ kN-m

Chapter 13

13.2. $M_{AB} = -98$ kN-m; $M_{BA} = 74$ kN-m; $M_{BC} = -74$ kN-m; $M_{CB} = 38$ kN-m
13.4. $M_{BA} = 108.9$ kN-m; $M_{BC} = -108.9$ kN-m; $M_{CB} = 42.2$ kN-m; $M_{CD} = -42.2$ kN-m
13.6. $M_{AB} = -8$k-ft; $M_{BA} = 14$k-ft; $M_{BC} = 2$k-ft
13.8. $M_{AB} = 62.2$ k-ft; $M_{BA} = 124.4$ k-ft; $M_{BC} = -124.4$ k-ft; $M_{CB} = 239.5$ k-ft;
$M_{CE} = -203.7$ k-ft; $M_{CD} = -35.8$ k-ft   **3.10.** $M_{BA} = 44.4$ k-ft; $M_{BC} = -44.4$ k-ft
13.12. $M_{BA} = 144$ kN-m; $M_{BC} = -144$ kN-m; $M_{CB} = 96$ kN-m; $M_{CD} = -96$ kN-m

Chapter 14

14.2. $M_{AB} = -6.9$ kN-m; $M_{BA} = +92.9$ kN-m; $M_{BC} = -92.9$ kN-m; $M_{CB} = +45.4$ kN-m;
$M_{CD} = -45.4$ kN-m; $M_{DC} = -22.7$ kN-m
14.4. $M_{BA} = 138.0$ k-ft; $M_{BC} = -138.0$ k-ft; $M_{CB} = 77.5$ k-ft; $M_{CD} = -77.5$ k-ft;
$M_{DC} = 18.8$ k-ft; $M_{DE} = 18.8$ k-ft
14.6. $M_{AB} = 4.6$ k-ft; $M_{BA} = 9.2$ k-ft; $M_{BC} = 18.5$ k-ft; $M_{BD} = 12.3$ k-ft; $M_{DB} = 6.2$ k-ft
14.8. $M_{BA} = 81.8$ kN-m; $M_{BC} = -99.9$ kN-m; $M_{CB} = 86.6$ kN-m; $M_{CD} = -50.9$ kN-m;
$M_{DC} = 75.0$ kN-m; $M_{BF} = 18.1$ kN-m; $M_{FB} = 9.0$ kN-m; $M_{CG} = -35.7$ kN-m;
$M_{GC} = -17.9$ kN-m
14.10. $M_{BA} = 144$ kN-m; $M_{BC} = -144$ kN-m; $M_{CB} = 96$ kN-m; $M_{CD} = -96$ kN-m

Chapter 16

16.2.

$$K = AE \begin{bmatrix} .0256 & -.0192 & -.0256 & .0192 & 0 & 0 \\ -.0192 & .0811 & .0192 & -.0144 & 0 & -.0667 \\ -.0256 & .0192 & .0756 & -.0192 & -.05 & 0 \\ .0192 & -.0144 & -.0192 & .0144 & 0 & 0 \\ 0 & 0 & -.05 & 0 & .05 & 0 \\ 0 & -.0667 & 0 & 0 & 0 & .0667 \end{bmatrix}$$

16.4. $q_1 = -7.14$ k; $q_2 = -22.2$ k   **16.6.** $q_1 = 26.5$ kN; $q_2 = 43.0$ kN; $q_3 = 22.5$ kN
16.8. $q_1 = 18.1$ k-ft; $q_2 = 36.2$ k-ft; $q_3 = 13.8$ k-ft; $q_4 = 0.4$ k-ft; $q_5 = -17.4$ k-ft;
$q_6 = -8.7$ k-ft; $q_7 = -13.0$ k-ft; $q_8 = -6.5$ k-ft
16.10. $q_1 = -85.7$ kN-m; $q_2 = 68.6$ kN-m; $q_3 = -68.6$ kN-m

Index

A

Arches:
    analysis, 181-83
    definition, 4, 180
Axial deformation, 105

B

Beams:
    definition, 6, 78
    deflections, 112-20, 124-27, 142-45
    differential equation, 105
    influence lines, 161-67
    internal forces, 79
    moment diagrams, 80-91
    shear diagrams, 80-91
    types, 78
Bending moment diagrams:
    beams, 80-91
    frames, 91-98
    sign convention, 80
Bridge live loads, 13
Building live loads, 13

C

Cables, 3, 183-84
Cantilever structures, 30-31
Carry-over factor, 262
Castigliano's theorems, 220-23
Conjugate-beam method, 120-27
Conservation of energy principle, 136

D

Dead loads, 12
Deflections:
    beams, 112-20, 124-27, 142-45

Deflections: *(cont.)*
  conjugate-beam method, 120-27
  direct integration, 106-10
  frames, 145-48
  moment-area method, 110-20
  real work, 138-39
  trusses, 149-52
  virtual work, 139-52
Deformation method (*See* Stiffness
  method)
Deformation transformation matrix, 309
Degree of indeterminacy, 193
Determinants, 338-40
Differential equation for beams, 105
Distribution factor, 260-62

E

Effective joint loads, 325-28
Elastic support, 203-4
Element flexibility matrix:
  axially loaded element, 288
  flexural element, 283
Element stiffness matrix:
  axially loaded element, 303-4
  flexural element, 304-7
Energy methods:
  analysis, 220-32
  deflection calculation, 136-52
Equilibrium equations, 20-21
External work, 138, 141

F

Finite-element method:
  flexibility method, 278-97
  stiffness method, 302-28
Fixed-end moment, 241-42, 260
Flexibility matrix:
  definition, 278-79
  element, 283
  structure, 285-87
Flexibility method, 278-97
Force method (*See* Flexibility method)
Force transformation matrix, 284-85
Form, structural, 2-8
Frames:
  moment diagram, 91-98

shear diagram, 91-98
sidesway, 252-53, 271-75

H

Hinges, internal, 27-30
History of structural engineering, 15-18

I

Indeterminacy, degree of, 193
Indeterminate structures:
  consistent-deformation method,
    192-214
  definition, 188, 192
  least-work method, 220-32
  matrix flexibility method, 278-97
  matrix stiffness method, 302-28
  moment-distribution method,
    258-75
  slope-deflection method, 238-53
Influence lines:
  beams, 161-67
  concentrated loads, 172
  definition, 160-61
  distributed loads, 173
  trusses, 167-72
Internal redundants, 212-14, 229-32
Internal work, 138-39, 141

J

Joints, method of, 39-45

L

Least-work method, 220-32
Live loads, 13-15
Loads:
  bridge, 13
  building, 13
  dead, 12
  live, 13-15
  snow, 14
  wind, 14

M

Material properties, 9-11
Matrix algebra:
    addition, 335
    inversion, 340-42
    multiplication, 335-37
    partitioning, 342-44
    subtraction, 335
    transposing, 337
Matrix method:
    flexibility method, 278-97
    stiffness method, 302-28
Maxwell's law of reciprocal deflections, 221-22
Moment-area method, 110-20
Moment-area theorems, 110-12
Moment diagrams (*See* Bending-moment diagrams)
Moment-distribution method, 258-75

N

Nodes, 282

R

Reactions, 20-31
Reciprocal deformations, law of, 221-22
Redundants, 193
Rigid frames (*See* Frames)
Rings, analysis of, 230-32

S

Sections, method of, 45-48
Settlement of supports, 204-6
Shear deformation, 106
Shear diagrams:
    beams, 80-91
    frames, 91-98
    sign convention, 80
Sidesway:
    moment-distribution method, 271-75
    slope-deflection method, 251-53
Sign convention:
    bending moment, 80

moment-distribution method, 260
    shear, 80
    slope-deflection method, 239
    trusses, 39
Singularity functions, 109-10
Slope-deflection equation, 238-44
Slope-deflection method, 238-53
Snow loads, 14
Space trusses, 66-73
Stability and determinacy, 22-24
Stiffness matrix:
    definition, 302-3
    element, 303-7
    structure, 307-10, 310-14, 314-17
Stiffness method, 302-28
Strain energy, 136
Superposition, principle of, 118-20
Support settlement, 204-6
Support tyeps, 21-22
Symmetry, 230

T

Torsional deformation, 106
Trusses:
    analysis, 36-59
    assumptions, simplifying, 36-38
    definition, 5, 36
    deflections, 149-52
    determinate, 39-59
    indeterminate, 229-30
    influence lines, 167-72
    matrix analysis, 314-22
    method of joints, 39-45
    method of sections, 45-48
    sign convention, 39
    space trusses, 66-73
    stability and determinacy, 51-53

V

Virtual work, 139-41

W

Wind loads, 14
Work, 136-37